Signals and Communication Technology

This series is devoted to fundamentals and applications of modern methods of signal processing and cutting-edge communication technologies. The main topics are information and signal theory, acoustical signal processing, image processing and multimedia systems, mobile and wireless communications, and computer and communication networks. Volumes in the series address researchers in academia and industrial R&D departments. The series is application-oriented. The level of presentation of each individual volume, however, depends on the subject and can range from practical to scientific.

Indexing: All books in "Signals and Communication Technology" are indexed by Scopus and zbMATH

For general information about this book series, comments or suggestions, please contact Mary James at mary.james@springer.com or Ramesh Nath Premnath at ramesh.premnath@springer.com.

P. Mary Jeyanthi
Editor

Distributed Ledger Technology in Communication

Integration of IoT, Blockchain and Metadata

Springer

Editor
P. Mary Jeyanthi
School of Management
Presidency University
Bengaluru, Karnataka, India

ISSN 1860-4862 ISSN 1860-4870 (electronic)
Signals and Communication Technology
ISBN 978-3-032-07681-6 ISBN 978-3-032-07682-3 (eBook)
https://doi.org/10.1007/978-3-032-07682-3

This Springer imprint is published by the registered company Springer Nature Switzerland AG
The registered company address is: Gewerbestrasse 11, 6330 Cham, Switzerland

Preface

The emergence of **Distributed Ledger Technology (DLT)** has profoundly reshaped the way data, communication, and transactions are handled in the digital era. With the rapid proliferation of the **Internet of Things (IoT)** and the ever-expanding importance of metadata, the integration of blockchain technologies with communication systems has become both an opportunity and a necessity. This book, *Distributed Ledger Technology in Communication: Integration of IoT, Blockchain and Metadata*, is a collective effort to explore the synergies, challenges, and real-world applications of these transformative technologies.

The motivation for this book stems from the growing recognition that communication systems of the future must be secure, transparent, and scalable. While IoT devices generate massive volumes of data, metadata provides structure and meaning, and blockchain ensures trust, immutability, and decentralization. Together, these technologies promise to revolutionize domains ranging from healthcare and energy systems to supply chain management, smart tourism, and finance.

This book is organized into eleven chapters, each offering unique insights:

- The opening chapters establish the foundational concepts and frameworks, discussing metadata-driven blockchain solutions and the prospects of IoT integration.
- Subsequent chapters explore into platforms, enablers, and domain-specific applications, including healthcare communication, airport baggage management, and renewable energy optimization.
- Later chapters highlight the role of IoT, blockchain, and metadata in smart ecosystems, supply chains, and financial systems, with an eye on emerging challenges such as quantum resilience and AI-driven disruptions.

The chapters are contributed by **researchers, practitioners, and thought leaders**, ensuring a balanced perspective between theory and practice. Collectively, these works demonstrate that the integration of IoT, blockchain, and metadata is not a distant vision but a rapidly materializing reality with profound implications for industries and societies.

This book is intended for **research scholars, industry professionals, policy-makers, and students** in computer science, information systems, communication engineering, and business domains. It provides both a conceptual framework and practical case studies, making it a valuable resource for anyone seeking to understand or implement DLT-enabled communication systems.

I gratefully acknowledge the contributions of all authors, reviewers, and collaborators who shared their expertise to bring this book to fruition. We also extend our thanks to **Springer** for their support and guidance throughout the editorial process.

I hope that this book will serve as both a reference and inspiration for future explorations in the integration of IoT, blockchain, and metadata, driving forward innovations in secure, intelligent, and connected communication systems.

Bengaluru, India Dr. P. Mary Jeyanthi

About the Book

The rapid evolution of digital ecosystems has fundamentally transformed how data is generated, shared, and secured across industries. Distributed Ledger Technology (DLT)—popularly known through blockchain—has emerged as a cornerstone for enabling decentralized trust, transparency, and resilience in modern communication systems. When synergistically combined with Internet of Things (IoT) and **metadata-driven intelligence**, DLT unlocks powerful capabilities for automation, interoperability, and secure data exchange.

This edited volume presents a comprehensive exploration of the convergence of **DLT, IoT, metadata, artificial intelligence, and platform** technologies across diverse application domains. The chapters bring together theoretical foundations, architectural frameworks, real-world case studies, and future-oriented perspectives that collectively address contemporary challenges in communication, security, scalability, and governance.

The book covers a wide spectrum of sectors including **healthcare, smart tourism, supply chain management, energy systems, airport operations, finance, and cloud-driven platforms**, demonstrating how decentralized intelligence is reshaping enterprise and public infrastructures. Special emphasis is placed on emerging paradigms such as **smart contracts, machine-to-machine communication, tokenization, AI-driven decision systems, and quantum-resilient auditing**.

Designed for **researchers, postgraduate students, academicians, policy makers, technology leaders, and industry practitioners**, this volume serves both as a scholarly reference and a practical guide. By bridging research innovation with applied use cases, the book contributes meaningfully to the global discourse on decentralized digital transformation and intelligent communication systems.

Contents

About the Editor

Dr. P. Mary Jeyanthi is a distinguished AI/ML & business analytics professor, researcher, and digital transformation strategist with extensive expertise in Distributed Ledger Technology, Internet of Things, metadata analytics, artificial intelligence, and decision intelligence systems. She has authored and edited several academic books and peer-reviewed research publications, with her work indcxed in leading international databases.

Dr. Jeyanthi has played a significant role in mentoring research scholars, guiding industry-aligned academic initiatives, and contributing to interdisciplinary research at the intersection of analytics, emerging technologies, and enterprise systems. Her academic and professional engagements span higher education, research leadership, consulting, and policy-oriented innovation.

As the editor of this volume, she brings together diverse scholarly voices to present a cohesive and forward-looking perspective on decentralized and intelligent communication ecosystems, making this book a valuable contribution to both academia and industry.

Contributors

Usha Badhera Department of Business Analytics, Jaipuria Institute of Management, Jaipur, Rajasthan, India

M. Banurekha Dr.N.G.P.Arts and Science College, Coimbatore, Tamilnadu, India

Saurabh Chandra School of Law, Bennett University, Greater Noida, India

Violeta Cvetkoska Ss. Cyril and Methodius University in Skopje, Faculty of Economics—Skopje, Skopje, North Macedonia

Ljupcho Eftimov Ss. Cyril and Methodius University in Skopje, Faculty of Economics—Skopje, Skopje, North Macedonia

Srikant Gupta Department of Operations and Decision Sciences, Jaipuria Institute of Management, Jaipur, Rajasthan, India

Kirti Jainani Professor (Marketing), Faculty of Management Studies, Institute of Rural Management (FMS-IRM), Jaipur, Rajasthan, India

Pankaj Jainani Principal Artitect and AVP, Genpact India Pvt. Ltd., Bangalore, Karnataka, India

P. Mary Jeyanthi School of Management, Presidency University, Bengaluru, Karnataka, India

Deepika Joshi St. Joseph's Institute of Management, Bengaluru, India

Christian Kaunert Dublin City University, Dublin, Ireland; University of South Wales, Cardiff, UK

Bojan Kitanovikj Ss. Cyril and Methodius University in Skopje, Faculty of Economics—Skopje, Skopje, North Macedonia

Pooja S. Kushwaha Department of Business Analytics, Jaipuria Institute of Management, Indore, Madhya Pradesh, India

Devvandya Mishra Jaipuria Institute of Management, Jaipur, Rajasthan, India

Neelam Naik SVKM's Usha Pravin Gandhi College of Arts, Science and Commerce, Mumbai, India

R. Oum Kumari Jaipuria Institute of Management, Jaipur, India

Siddharth Pareek ICICI AMC, Connaught Place, New Delhi, Delhi, India

Harshali Patil Somaiya School of Basic and Applied Sciences, Somaiya Vidyavihar University, Mumbai, India

Kiril Postolov Ss. Cyril and Methodius University in Skopje, Faculty of Economics—Skopje, Skopje, North Macedonia

Gaurav Sharma Jaipuria Institute of Management, Jaipur, India

Bhupinder Singh Sharda University, Greater Noida, India

R. Vedapradha St. Joseph's College of Commerce, Bengaluru, India

Lokesh Vijayvargy Jaipuria Institute of Management, Jaipur, Rajasthan, India

Chapter 1
Introduction: The Convergence Era: Decentralization Meets Intelligence

P. Mary Jeyanthi

Abstract The convergence of Distributed Ledger Technology (DLT), Internet of Things (IoT), and metadata marks a paradigm shift in digital communication and operational ecosystems. As industries transition toward Industry 5.0 and Society 5.0, the demand for secure, transparent, and interoperable communication systems has become indispensable. IoT, with its capability to generate vast streams of real-time data, has transformed the way organizations collect and utilize information, while blockchain introduces trust, immutability, and decentralization into these processes. Metadata, acting as the contextual backbone, enables meaningful interpretation and intelligent decision-making, creating an integrated framework for next-generation communication systems. This chapter provides a foundational understanding of the synergy between these technologies, emphasizing their role in enabling trusted machine-to-machine communication, automation, and digital resilience across sectors such as health care, energy, logistics, finance, and tourism. It addresses the pressing challenges of interoperability, privacy, scalability, and standardization, while highlighting the transformative potential of combining blockchain's security features with IoT's connectivity and metadata's contextual intelligence. By setting the stage for the subsequent chapters, this introduction underscores the strategic importance of adopting decentralized, intelligent infrastructures for sustainable and scalable solutions. It not only explores the theoretical underpinnings of this integration but also outlines the practical implications for building futuristic ecosystems capable of supporting smart contracts, real-time analytics, and autonomous coordination. Ultimately, this convergence paves the way for a digitally empowered society where trust, transparency, and efficiency become foundational principles of communication and collaboration.

Keywords Distributed ledger technology (DLT) · Internet of things (IoT) · Metadata · Blockchain integration · Decentralized communication

P. M. Jeyanthi (✉)
School of Management, Presidency University, Bengaluru, Karnataka, India
e-mail: Dr.maryprem@gmail.com

P. M. Jeyanthi (ed.), *Distributed Ledger Technology in Communication*, Signals and Communication Technology, https://doi.org/10.1007/978-3-032-07682-3_1

1.1 Introduction

In the age of ubiquitous connectivity, the fusion of Distributed Ledger Technology (DLT), Internet of Things (IoT), and metadata is revolutionizing how communication systems function across sectors. The emergence of blockchain as a trustless, decentralized system, when integrated with smart IoT infrastructures and enriched metadata, offers transformative potential for real-time communication, enhanced security, transparent operations, and intelligent decision-making.

This book, "Distributed Ledger Technology in Communication—IoT and Blockchain Integration," aims to bring together cutting-edge research, practical use cases, and futuristic outlooks on how blockchain-enabled communication ecosystems can be built through metadata-driven IoT applications. As organizations and governments transition toward Industry 5.0 and Society 5.0, a coordinated digital infrastructure with secure and interoperable communication mechanisms becomes a necessity rather than a choice.

In the age of digital acceleration and hyperconnectivity, the convergence of Distributed Ledger Technology (DLT), Internet of Things (IoT), and metadata is redefining the foundational architecture of communication, collaboration, and coordination across diverse industries. As we move toward a data-centric society, the challenges surrounding data integrity, interoperability, transparency, and security have become critical concerns for technologists, policy-makers, and business leaders alike. This book aims to offer a comprehensive exploration of how blockchain and IoT—augmented by metadata frameworks—are not only reshaping traditional communication systems but also catalyzing innovation across verticals such as health care, logistics, finance, energy, and tourism.

At its core, this book presents a compelling narrative of how decentralized technologies, when effectively integrated with intelligent sensing and contextual metadata, create new avenues for seamless, secure, and sustainable digital communication. It captures the essence of next-generation technological ecosystems by offering theoretical perspectives, practical insights, application case studies, and future trajectories.

1.2 The Need for Integrated Solutions

Over the past decade, both IoT and blockchain have emerged as transformative technologies, yet their evolution has been largely independent. While IoT has proliferated into billions of devices generating real-time data across sectors, blockchain has built a reputation for enabling trust, decentralization, and immutability in digital transactions. However, the integration of these technologies with metadata—contextual information that gives meaning to raw data—presents a unique opportunity to revolutionize how machines communicate, decisions are made, and operations are managed.

This integration addresses three fundamental challenges in communication:

- Data Authenticity and Provenance—Blockchain ensures that the data generated by IoT devices is tamper-proof and auditable.
- Contextual Intelligence—Metadata provides essential context for real-time decision-making and automation.
- Secure, Decentralized Coordination—Smart contracts and blockchain networks allow machines and stakeholders to engage in trusted, automated interactions without intermediaries.

1.3 Scope of the Book

The objective of this book is to delve deep into the theoretical foundations, implementation frameworks, domain-specific use cases, and emerging challenges related to the convergence of DLT, IoT, and metadata. It is designed for researchers, practitioners, policy experts, and graduate students who seek to understand and implement these technologies in real-world scenarios.

The chapters in this book are thoughtfully arranged to follow a progressive learning path—starting from fundamental integration concepts to sectoral applications, and finally, toward future-ready innovations in cross-industry domains.

1.4 Chapter Contributions and Discussions

Chapter 1: Introduction

This chapter sets the foundation of the book by highlighting the convergence of Distributed Ledger Technology (DLT), Internet of Things (IoT), and metadata in the modern communication landscape. It outlines the significance of secure, scalable, and interoperable systems and introduces the key sectors (health care, energy, finance, logistics, tourism, etc.) where this integration can bring disruptive innovation. The chapter also briefly discusses challenges such as interoperability, standardization, and privacy, laying the groundwork for the chapters to follow.

Chapter 2: A New Horizon for Metadata: Blockchain-Driven Data Management Solutions

Metadata plays a crucial role in data governance, traceability, and automation. This chapter explores how metadata, when stored and processed via blockchain, can be used to facilitate tamper-proof data provenance, enhance data lineage, and enable secure smart contracts. It delves into applications such as digital identity management, document verification, and regulatory compliance, proposing that blockchain serves as the backbone of metadata integrity in large-scale IoT communication networks.

Chapter 3: Blockchain and IOT Integration: Prospects and Obstacles

Chapter 3 examines the synergies and real-world integration challenges between IoT and blockchain. It discusses how IoT devices collect real-time data, and how blockchain ensures its security, integrity, and decentralization. Obstacles such as latency, scalability, energy consumption, and storage limitations are critically analyzed. The chapter also outlines technical architectures and middleware solutions to enable scalable blockchain-IoT systems, making it a vital contribution to the book's technology discourse.

Chapter 4: Cloud Driven Platform Services as an Integration Enabler for DLT (Blockchain), IoT and Metadata

This chapter positions cloud computing as the bridge that connects the scalability of IoT with the immutability of blockchain. It presents cloud-based middleware and APIs that enable real-time data flow between IoT devices and blockchain networks. Discussions include hybrid cloud-edge models, blockchain-as-a-service (BaaS), and security frameworks for distributed cloud ecosystems. This is especially relevant for enterprises looking for cost-effective, scalable integration architectures.

Chapter 5: Absorption of Metadata, IoT, and Blockchain for Security Smart Contacts Communication between Hospital Industry and Patients

Focusing on health care, this chapter demonstrates how smart contracts and metadata can enhance secure communication between hospitals and patients. It explores real-time patient monitoring, data sharing with consent, and tamper-proof medical records using blockchain. Use cases include IoT-based health wearables and biometric identity management. The chapter shows how this integration can uphold privacy, reduce fraud, and ensure seamless hospital workflows.

Chapter 6: Blockchain Technology Enabled Baggage Management at Airport

This chapter applies blockchain and IoT in aviation, particularly airport logistics. It addresses the long-standing issue of lost or delayed baggage by proposing blockchain-enabled asset tracking. IoT sensors track baggage movement, while blockchain records every transaction, ensuring visibility and reducing disputes. The chapter includes a model architecture and cost–benefit analysis for airline operators, demonstrating how this integration can improve passenger experience and operational efficiency.

Chapter 7: Optimizing Solar Energy Systems through IoT, Metadata, and Blockchain Technologies

Focusing on the renewable energy sector, this chapter explores how IoT-powered solar panels, when combined with blockchain smart contracts and rich metadata, can enable automated energy distribution, dynamic pricing, and carbon credit trading. It explains peer-to-peer energy exchanges and grid balancing in smart cities. The blockchain ledger helps in maintaining trust, while metadata provides insights for energy optimization, predictive maintenance, and compliance.

Chapter 8: Smart Tourism Ecosystems: Integrating IoT, Blockchain and Metadata for a Connected Future

This chapter envisions tourism ecosystems powered by IoT sensors, enriched traveler metadata, and blockchain-driven smart services. From digital identity and secure e-visas to personalized recommendations and tamper-proof transaction records, the chapter outlines how this integration enhances safety, personalization, and mobility. It emphasizes traveler trust, contactless experiences, and data-driven governance in building smart tourism destinations.

Chapter 9: Disruptive Technologies Reshaping Supply Chain Management for the Future: AI Integration, IoT, and Blockchain

This chapter builds a strong theoretical and practical foundation for next-generation supply chain systems. It outlines how blockchain provides transparency, IoT ensures real-time tracking, and AI enables predictive decision-making. It discusses inventory control, automated replenishment, fraud detection, and demand forecasting in digitally transformed supply chains. The use of AI models trained on IoT-generated metadata presents a futuristic look into self-learning logistics systems.

Chapter 10: Supply Chain Tracking and Management with Blockchain

A focused extension of Chapter 9, this chapter offers detailed case studies and technical blueprints for blockchain-based supply chain tracking. It highlights use cases in pharmaceuticals, agriculture, electronics, and luxury goods, showcasing the end-to-end tracking of products using decentralized systems. Smart contracts automate approvals and payments. The discussion extends to GS1 standards and blockchain interoperability for global supply networks.

Chapter 11: Blockchain-Enabled Cross-Asset Finance: Tokenization, AI-Driven Trading and Quantum-Resilient Auditing

The final chapter shifts the lens to financial services, presenting how blockchain enables tokenization of physical and digital assets, AI-driven autonomous trading, and quantum-resilient cryptographic audits. It discusses how financial metadata from IoT devices (e.g., insured goods, real estate, energy credits) can be tokenized for decentralized finance (DeFi) platforms. The chapter highlights the future of secure, interoperable, and intelligent financial ecosystems.

1.5 Interdisciplinary Relevance

The book is highly interdisciplinary, merging concepts from computer science, engineering, business, and social systems. It not only explains the technical mechanics but also considers legal, regulatory, and ethical implications of adopting these disruptive technologies.

Whether it's in enabling smart hospitals, decentralized energy grids, intelligent logistics, or autonomous financial systems, the chapters demonstrate how communication and coordination are being redefined through decentralized intelligence and machine-to-machine trust.

1.6 Conclusion and Reader Takeaway

This book is a roadmap for navigating the future of digitally distributed communication. It shows that when IoT-generated data is enriched with metadata and protected by blockchain, we unlock a powerful triad that drives innovation, transparency, and trust.

Readers will gain

- A solid understanding of DLT and IoT integration frameworks.
- Practical case studies across industries.
- Insights into emerging technologies like smart contracts, AI, cloud platforms, and quantum security.
- A forward-looking perspective on how to leverage these technologies for sustainable development, operational excellence, and digital resilience.

As global systems increasingly become connected, automated, and decentralized, the insights offered in this book will be indispensable for shaping strategies, architectures, and innovations that define the communication networks of tomorrow. This compilation offers an interdisciplinary perspective—from health care and energy to supply chain, finance, tourism, and aviation—revealing how integrated DLT systems can reshape verticals and provide resilience, automation, and intelligence.

Chapter 2
A New Horizon for Metadata: Blockchain-Driven Data Management Solutions

Neelam Naik and Harshali Patil

Abstract Metadata means "data about data," and it is essential for the efficient management and recovery of data. It improves the understanding and usefulness of the original information. According to Gartner, metadata turns information into an asset. Metadata significance is associated with its ability to provide the content with relevant context. It helps organizations to quickly and resourcefully undertake impactful decisions and extract thoughtful insights from their assets effective metadata management is a basis for digital asset management, which ensures the sustained accessibility and relevance of digital content over time. In blockchain technology, metadata indicates the additional data that provides context about transactions or blocks without changing the core data. Metadata on blockchains is immutable and secure; it also ensures that it cannot be modified once added to the block, thus it preserves the reliability of the information. It includes information such as transaction timestamps, contract terms, and user identities with improved transparency as well as traceability. Metadata is extremely important in applications such as smart contracts, digital identity verification, and supply chain management because it adds layers of information to complicated processes and agreements. Blockchain is the most promising technology for future technological development. Metadata on blockchain will help to provide different perspectives on existing systems. To describe effective metadata management, two blockchain-based systems: the Public Distribution System (PDS) and the Blood Bank System (BBS) are discussed in this chapter. Metadata management on blockchain in the PDS in India can revamp the distribution of food grains, and essential commodities to desirable beneficiaries. It can improve transparency, reduce fraud, and enhance the efficiency of the distribution system. The traditional BBS has limitations related to detailed information about blood, which leads to challenges in managing blood quality, supply, and demand.

N. Naik
SVKM's Usha Pravin Gandhi College of Arts, Science and Commerce, Mumbai, India
e-mail: neelam.naik@upgcm.ac.in

H. Patil (✉)
Somaiya School of Basic and Applied Sciences, Somaiya Vidyavihar University, Mumbai, India
e-mail: harshali.yp@somaiya.edu

P. M. Jeyanthi (ed.), *Distributed Ledger Technology in Communication*, Signals and Communication Technology, https://doi.org/10.1007/978-3-032-07682-3_2

The blood bank system requires careful planning, adherence to healthcare regulations, and collaboration with stakeholders. Metadata management in the blood bank system can become a boon to the healthcare sector by overcoming the challenge of blood quality.

Keywords Blockchain · Smart contracts · Metadata management · Public distribution system · Blood bank system · Interoperability · Hybrid system

2.1 Introduction

2.1.1 Brief Overview of Traditional Metadata Management

Metadata is essential because it maximizes the value of data assets through context and consistent use of terminology. Metadata is the data in the context of who has created it, what is its level of privacy, where it is stored, how it is stored, why it has a specific format, and when it was created. Most organizations need metadata management for handling their increased data management practices due to the need for data governance, data regulation, and data compliance requirements. In the business domain, insights from the data can be obtained with the help of high-quality data and by applying analytics to it. Along with the traditional data collection sources, the data collection becomes complex; hence, to handle the complexity, metadata management becomes a crucial part. To fulfill the growing demand and to accelerate the transformation initiatives, such as data digitization and multi-modal access of data, metadata management plays an important role.

Meta management has a capability that includes data inventory, which automatically solves data ambiguities and identifies the attribute relations and connections with available data sources. With the help of user tagging and rating, data enrichment could be possible. A historical record of data and its origin can be determined using metadata management. Also, if there are any changes in the metadata, the impact analysis can be used to find the long-term effect of this change. The dynamic metadata management's extensive use can result in automation through artificial intelligence. A variety of users, such as data architects, data engineers, data administrators, and data analysts, can get the required information through collaborative methods. To improve the data quality in workflows, metadata creators and consumers contribute collaboratively. Metadata management can assist privacy and security with the help of policy management and regulations. The right metadata management can lead to data intelligence. It could be possible by creating a repository of enterprise data assets with the principles of data administration, confidentiality, data catalog, and history of data. Implementation of Metadata management involves three levels such as metadata capture, metadata storage, and administration. It is important to understand the relationship between these three levels. Terms represent the common business language and its definitions, for example, policy manuals, contracts, or industry standards.

Attributes represent business resources such as systems or reports, which include documentation, data dictionaries, and data models. Elements represent database tables or reports. It includes data models, database catalogs, and spreadsheets. To extract quality data, enterprise-level policy management is required. Increasing the business value, to enhance creativity, and reduction of risk can be possible through metadata management. Metadata is like a smartphone box where the product provides information about itself. From a data perspective, it improves the utility of the product and its packing despite not directly affecting the performance of the smartphone [1].

2.1.2 Introduction to Blockchain Technology and Its Relevance to Data Management

A sophisticated database system that enables transparent information exchange between business network members is referred to as blockchain technology. This data is stored in the form of blocks, and these blocks are interconnected in a chain, which ensures accuracy in order of time. As the data is in chronological order, the data cannot be removed or altered without having approval from the network. This feature enables us to track payments, orders, and transactions while generating an irreversible or immutable ledger employing blockchain technology. Blockchain technology includes security features that prevent unauthorized transaction entries and maintain consistency in the shared view of these transactions. Figure 2.1 represents key features of blockchain technology.

Decentralization: Blockchain distributes data across the network, which reduces the risk of failures and attacks, whereas in conventional databases, the central authority has control of data.

Fig. 2.1 Key features of blockchain technology

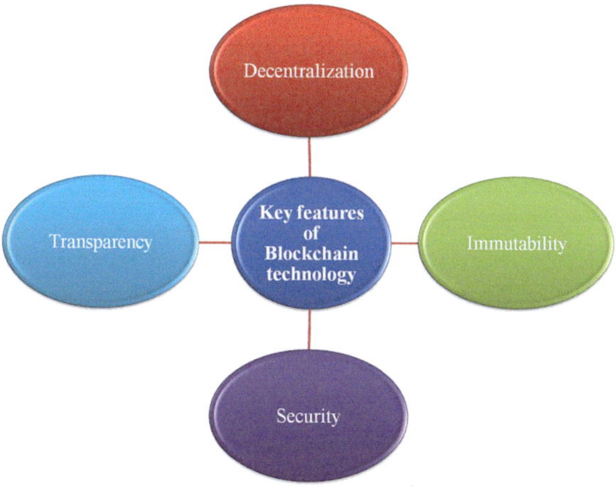

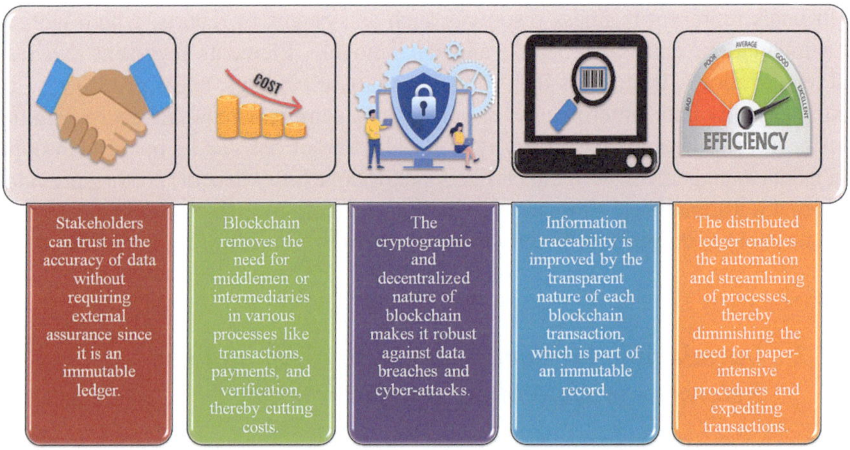

Fig. 2.2 Advantages of blockchain for data management

Transparency: The blockchain allows all participants to see any alterations made to the blockchain and requires consensus before being officially documented. Thus, it ensures a high level of transparency.

Security: The blockchain protects data from tampering and fraud with the help of cryptographic algorithms applied to data.

Immutability: Once a transaction has been recorded, it is impossible to change the data in any block without altering all the blocks that come after it. This process requires the consensus from the majority of the network.

In Spite of features like decentralization, transparency, security, and immutability, the challenges the blockchain faces are scalability, adaptation to new technology, approval by regulators, and use of energy. Blockchain provides a revolutionary approach to data management by enhancing its key features. Figure 2.2 represents the advantages of blockchain technology for data management.

2.1.3 The Basics of Metadata and Blockchain

2.1.3.1 Explanation of Metadata and Its Importance in Digital Ecosystems

Metadata is data about data. In blockchain technology, metadata refers to data that provides additional data about transactions, and it does not hamper the actual data of a transaction. The metadata can be used to improve usability and functionality. It provides details such as the origin, history, and actual data. Metadata enhances

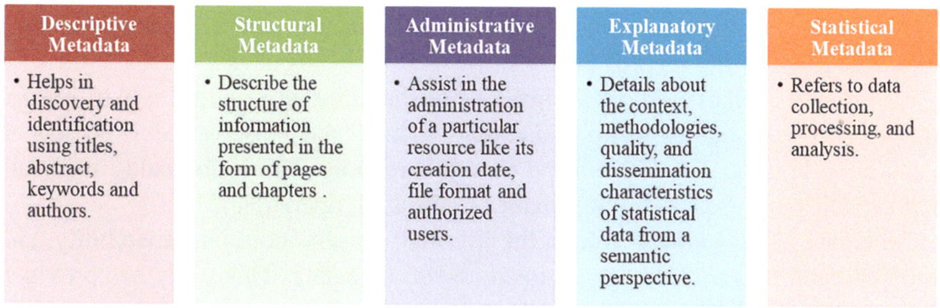

Fig. 2.3 Categories of metadata on the basis of usage

the user interface and experience by comprising tags or labels, making the transaction more understandable for the user. Metadata also helps to improve interoperability between blockchain systems and external systems and provides insights into blockchain activities. Metadata is stored in the blocks of the blockchain using the smart contract. It is permanently stored on the blockchain once the transaction is broadcast. The metadata cannot be deleted as blockchain has features as immutability and irreversibility. Metadata is divided into two types based on data storage capability.

On-chain metadata: It is publicly available and can be viewed by anyone with access to the blockchain. In this, metadata is stored on the blockchain, hence known as on-chain metadata, which consists of information like labels, comments, and references to other contracts.

Off-chain metadata: It is included in the transaction without being stored directly on the blockchain. The information is referred to within a transaction through links or external references directing to additional documents, resources, and records that are available online. As the off-chain metadata is not directly included in the transaction processing, it is used in elaborative metadata.

Based on the usage of metadata, it can be categorized into five types as shown in Fig. 2.3.

Metadata, within the blockchain, serves as an asset for enhancing functionality and expanding the scope of transactions while upholding the authenticity and safety of the core data.

2.2 Literature Review

Metadata management includes understanding data items that are created and utilized during the project life cycle. In the geoscience discipline, metadata is created and reused across many records generated throughout the life cycle of environmental research, encompassing the project's geoscience datasets and data collection sites. Metadata lifecycle methods include the collection, sharing, and discovery of data.

Metadata construction and administration can be updated and shortened by reusing metadata across many records [1].

Crossref is a charitable open digital infrastructure that provides a platform for the research community to find, site, assess, link, and reuse research objects. It has around high-quality 100-million metadata records. This high-quality metadata handles millions of queries and provides transparent operation [2].

Metadata plays a crucial role in the life-long preservation, interoperability, and reuse of data. It is a challenge to create useful metadata. This study has provided a personalized solution to create a domain-based metadata structure by involving experts and researchers in this process [3].

This study has provided a music metadata copyright protection method using blockchain technology to identify suspected plagiarism in music tracks. Using the smart contract technology, metadata such as lyrics, rightful producer details, time signatures, and melodies can be attached to the music track. This method will help to reduce the number of copyright violations in the music industry [4].

The concept of pidgin metadata is proposed to achieve reliable metadata interoperability using a three-layered metadata framework. The first layer is responsible for serving mediator for semantic association between metadata elements. Layer two is made up of metadata elopements to describe similar aspects of information resources. The framework for the resource description is maintained by the third layer [5].

The authors of the research paper [6] have proposed a private Ethereum blockchain-based solution to program blood donation management to make this system traceable, secure, auditable, decentralized, and transparent. This system provides non-critical and large data off-chain using the distributed storage of the InterPlanetary File System (IPFS) [6]

The exploratory study [7] explains the use of distributed ledger technologies within the plasma derivatives supply chain. The model has been developed using blockchain technology that verifies the origin of plasma and deals with the ethical and risk-related issues in the plasma supply chain [7].

In this paper [21], the authors have introduced the blood management process based on blockchain technology and provided Blood and Product-Chain, using a distributed ledger. It manages all related information about blood and its products. Hyperledger Fabric is used to implement the Blood and Product-Chain proof-of-concept and assess it in data creation and data access. Medical staff (nurses and doctors) gather the patients' records to learn about past treatments. The hospital treating the patients will receive an update with all this information. The hospital will file a request by updating the distributed ledger with the current blood amount if there is not enough blood or blood products during treatment. In Blood and Product-Chain, all relevant parties are informed about blood volume, blood type, and other unique requirements. The information about medical facilities is also updated at this stage. To maximize blood availability, blood volumes in a medical facility's system might be shared with other facilities. To accomplish that, the system must update the status and contact details of the transporter with the locations for the collection and receiving of blood and blood products. Subsequently, the transporter is required to consistently update the distributed ledger with the shipment's progress. The stakeholders could verify and moderate all information regarding location, time, and order. The hematological hospital receives this request. The process for determining how much blood and blood products is on hand is outlined in the next step. If the required quantity of blood cannot be obtained, then the system gets in touch with the donors to arrange for a blood test and donation. The candidates with the closest addresses are chosen from the list of donors that were collected from earlier blood donation events.

The research paper [22] suggests a blockchain-based system to enhance blood donation management. The present systems are confronted with problems like the lack of specific information on blood and the complications in handling blood quality and supply. Through BloodChain, which operates on Hyperledger Fabric, the problems of insecurity, privacy violation, and trust are solved by the traceability, transparency, and reliability of the blood donation supply chain. The suggested system is designed to furnish all-around data about blood source, quality, consumption, and disposal, thus, improving the general productivity and confidence in the management of blood. The main aspect of this research is the use of a private blockchain for B2B transactions, implementation based on Hyperledger Fabric architecture, enhanced traceability and information visibility, and the evaluation of the system in several situations to prove its effectiveness.

The research study [23] solves the problems of the blood banks, like the supply–demand imbalance and data privacy. It proposes a management platform comprising three modules: predicting the blood needs, determining the areas for blood donation drives, and ensuring the blood supply chain.

Through an analysis of 23 pertinent papers, this review [24] investigates the effectiveness of the PDS in achieving food and nutritional security in India. The results emphasize the PDS's contribution to the fight against hunger and malnutrition, but they also draw attention to the system's limited ability to improve food security and lower childhood mortality because of operational inefficiencies. The assessment highlights how PDS has the capacity to alleviate food hunger, provided that operational inefficiencies and environmental effects are sufficiently addressed through policy reforms [24].

2.3 Challenges in Traditional Metadata Management

Trust is a big deal when it comes to traditional metadata management systems. In these systems, usually, users' trust is a central authority to handle, update, and keep metadata secure. If there is a single authority existing, then chances of vulnerability may be there, where the data's integrity may be compromised either by accident or on purpose. Users may be hesitant to depend on such a single authority system without having a strong mechanism to verify that metadata is genuine and accurate. Such possibilities are especially important where data integrity is most important in domains like healthcare, law, and finance.

Security challenges can arise from both internal and external threats. There is always a chance, internally, that employees or system administrators may modify the data without sufficient oversight. Traditional systems are vulnerable to cyber-attacks. A variety of attacks, like data breaches, ransomware attacks, or hacking, can lead to stolen, corrupted metadata. If the data is centralized, then the risk is further increased as the data centralization causes a single point of failure, which can be targeted by the attacker.

Centralization indicates that all the metadata is controlled and stored in one place or managed by a single authority. With this, it becomes easier to manage and handle things, but it has drawbacks as well. One issue is that it makes the system less transparent and more susceptible to corruption and misuse. Also, with centralization, the delays in accessing and processing data will be there, and this reduces the overall efficiency of the responsiveness of the system.

2.3.1 The Complexity of Managing Metadata at Scale

Data Volume: As organizations collect and store more data, the number of metadata also skyrockets. Dealing with huge volumes of metadata can be quite a logistical headache. Traditional systems may have a hard time scaling up smoothly, resulting in higher costs and lower performance.

Diversity of data sources and data formats: Metadata management includes data from different sources and formats. These may include structured information, unstructured information, or documents. Usually structured indicates a standard format is applied, such as databases. Unstructured includes the data available in social media posts or in the form of emails. Managing such metadata is complex because each kind of data may need to be handled, stored, and protected differently.

Integration Challenges: Integration of metadata across various systems and platforms is challenging because different systems could have different formats for storage, management protocols, and metadata standards. Hence, it is challenging to aggregate or synthesize such data.

Regulatory Compliance: Organizations may need to implement several compliance methods at the same time since various roles may require specific handling, retention, protection, and deletion policies for metadata.

2.4 Blockchain as a Solution

Blockchain technology provides innovative solutions for the core issues in metadata management, which include decentralization, security, data integrity, traceability, and interoperability.

Decentralization: Blockchain uses a distributed ledger; it operates on a peer-to-peer network in which every participant has a copy of the ledger. It decreases the chances of the concentration of power and manipulation. The modifications of the metadata are done by the consensus mechanism of the blockchain. It implies that all the changes are confirmed by different people and that the unwanted modifications are prevented. In a nutshell, all the network members must reach a consensus before the metadata on the blockchain is changed.

Security: Blockchain protection of its data is based on the use of top-notch encryption methods. This makes the data safe from data theft and modifications. Since blockchain is a distributed system, there is no issue of single-point failure; thus, metadata management can be made more secure, also it protects the system against threats and hacking.

Data Integrity: Immutability is the most vital characteristic of the blockchain because every time data is recorded on a blockchain, it cannot be changed or erased without the agreement of the network. This guarantees data accuracy and data longevity. In blockchain, each entry is the one which is timestamped, and so, it creates a void-proof and traceable record wherever a piece of metadata is added or modified.

Traceability: Blockchain technology makes it possible to monitor the past of each piece of metadata by giving a clear audit trail of all the modifications and transactions. The provenance tracking is useful in cases when the source and the background of the data are the main concerns.

Interoperability: The blockchain can create the standards for metadata management among different systems and organizations, thus, it will be easier for interoperability and data sharing.

Blockchain solves the main problems in metadata management, and it is very suitable for different sectors, like financial services, healthcare, digital asset management, and domains where reliable metadata is needed.

The decentralized architecture of blockchain helps to enhance security by distributing data over several nodes, and it also requires consensus for any changes to be done in the data. This mechanism protects the data against fraud as well as cyber-attacks. It also establishes trust by overcoming the need for central access and ensuring transparency.

Enhanced Security: The traditional centralized system has an issue of a single point of control, which in turn can be the single point of failure, whereas in a decentralized system, data is distributed in different locations, such as different computers or servers in the network. This means that even if one or several nodes are compromised, the network can still work without any loss of information.

Decentralized Networks are more secure in case of Denial of Service (DoS) attacks or hacking. DoS problem is associated with the central server; usually attacker targets the server and floods it with traffic until it stops responding to legitimate users. In blockchain, there is no such central server to target, and that's why attackers must target a significant portion of the network to have control over it. But this is highly impractical as well as it requires many resources.

In a decentralized system, agreement by various parties is necessary before any changes can be made in the data. This means a single entity or an individual will never be able to alter the data. To alter any piece of information majority of the nodes must agree with the changes in data, which makes unauthorized alteration very difficult.

Increased Trust: Decentralization is often bundled with transparency, as each node in the network has an entire copy of a ledger. Hence, all transactions and their supporting data can be checked by anyone. This fosters trust among the users. Here user need not rely on any central authority to confirm the transaction. In conventional systems, banks, agents, and brokers act as intermediaries during transaction settlement. In a decentralized network, the intermediate trusted parties are eliminated. On a decentralized network, every transaction is documented on a public ledger. As it is open and can be reviewed by anyone, the trust is established through accountability.

Implications across industries: The traditional finance industry has been changed by decentralized blockchain. This enables transactions without the need for banks or clearing houses, reducing fraud while cutting down on the transaction cost. The decentralization technique will help to enhance the trust between suppliers, shippers, and consumers. This is possible through creating unchangeable records of the movements of goods, from source to destination. Decentralized apps can guarantee that several healthcare professionals have access to up-to-date and accurate information without compromising security.

Immutability Ensuring the Integrity of Metadata

Principle of immutability: One of the foundational features of metadata is immutability, which ensures integrity of metadata. In the blockchain, once the information is recorded, it becomes immutable, and any individual entity cannot modify or delete it. This is because each block of data, including its metadata relates to the previous block through a hash value. The hash is the unique, alphanumeric sequence that generates data present in the block.

Ensuring Metadata Integrity: In blockchain, every block contains the hash of the previous block. Changing any part of the data will change its hash. And as blocks are connected by these hashes, modifying one block would mean recalculating all others, which becomes almost computationally impossible in the centralized network. Proof-of-work (PoW) and Proof of Stake (PoS) are used for adding new information to the blockchain. These protocols among the network nodes decide whether the entries are valid or not, which helps to prevent unauthorized modification of metadata. Each node in the blockchain has a copy of the complete blockchain. Such distributed storage ensures that even if one part is compromised other nodes still have original data, which guarantees the integrity of metadata.

Benefits of metadata management: Metadata cannot be altered once it is integrated into the blockchain. Users can trust metadata accuracy. Any party applying the blockchain can independently confirm the origin and authenticity of metadata without knowing the other parties involved. The immutable metadata record supports legal and regulatory compliance and facilitates the auditing process. Every transaction has an unchangeable history that can be tracked by its initial source, along with any supporting metadata.

2.4.1 Advantages of Blockchain-Enabled Metadata Management

Enhanced Security and Reduced Risk of Tampering

Cryptography: The application of state-of-the-art cryptographic methods in blockchain provides data security; thus, any attempts to change the recorded information without being detected are automatically stopped.

Distributed Ledger: The blockchain ledger, which is kept on numerous nodes, makes it impossible to have a central point of failure. It, thus, greatly reduces the possibility of data tampering or loss.

Consensus mechanism: Consensus algorithms are of the essence in the blockchain for getting the network participants' agreement before the metadata can be modified thus, making the way for the protection of the blockchain, which means that the blockchain is now safe from any unauthorized changes.

The Traceability and Transparency of the Results for Auditing Purposes Are Enhanced

Immutable records: The records in the blockchain cannot be changed once they are inputted and therefore; they remain forever unaltered. The unchangeable feature ensures the existence of a solid and permanent record of all the transactions or the changes to metadata, which can be checked by anyone.

Audit Trail: The Audit Trail in the blockchain is the proof of every change made to the metadata which is documented and can be audited. Thus, with these traits, the audit process becomes obvious and makes adherence to the regulatory requirements possible, which in turn is a factor of accountability and transparency.

Real-time access: Blockchain offers instantaneous access to the stakeholders. It is a major factor that initiates access to the data and metadata from it. The on-the-spot access of the stakeholders to the data gives them the insights that they need. Besides, it also increases the follow-up of activities and the visibility of the process.

Increased Efficiency Through Smart Contracts and Automated Processes

Automation: Smart contracts automatically execute the transactions or data-related tasks, when a specific condition is satisfied. Thai reduces the intervention of humans and helps to speed up the process.

Reduced error: Through the automation of metadata management, blockchain helps to mitigate human errors such as data entry, mistakes, or data processing delays.

Cost efficiency: Blockchain technology reduces the reliance on intermediaries, lowers transaction costs, and streamlines operations by enabling peer-to-peer interactions on a distributed platform.

2.4.2 Challenges and Considerations

Blockchain technology has many problems in its performance and scalability, especially the proof-of-work mechanisms which may decrease the transaction speed and increase latency, which, in turn, makes it hard for it to handle large metadata volumes. As the blockchain networks grow, the data load becomes the main problem which will be the main reason for the scalability issues, in the systems where high throughput and storage are needed. Besides, the integration of blockchain with the current IT infrastructures is challenging and expensive hence it is time-consuming, to ensure the smooth deployment and operation of the system, the technical expertise and resources are required for the integration. The difficulties that researchers face show the necessity of constant technical improvements and careful integration strategies in the implementation of blockchain.

The transparency that is in the blockchain can be a threat to privacy needs, especially when it comes to managing sensitive metadata, therefore, it is very important to find a balance between the two. Even though blockchain technology offers some degree of pseudonymity, it might not be enough for every situation and hence, the use of extra privacy-enhancing tools like zero-knowledge proofs and private channels becomes a must. Besides, the use of high-tech encryption and data masking is the way to protect the privacy of the blockchain, but these methods need to be carefully designed to make the system work well and be easy to use. These strategies are the key to getting the most out of the blockchain, and at the same time, they are the tools to overcome privacy issues.

The global and decentralized nature of blockchain generates obstacles in the implementation of local data protection laws like GDPR which has its own set of rules and regulations for data handling. The lack of universal standards for blockchain technologies can lead to the incompatibility of systems which can be a reason for blockchain to be used by few. Furthermore, the quick growth of blockchain is often so much that it exceeds the current regulation, thus, it brings legal and regulatory problems that may be a reason for organizations not to use this technology. These issues point to the necessity to evolve the regulations and standards that will help ensure the safe and effective use of blockchain.

2.5 Case Studies and Applications

2.5.1 *Real-World Examples of Blockchain in Metadata Management Across Industries*

In various industries, blockchain is used as a decentralized and secure way of metadata management. Real estate, healthcare, finance, supply chain, and digital media are the major domains where blockchain is used for metadata management efficiently and transparently.

Digital Media: The digital content, such as images, videos, and documents, and their metadata are verified for authenticity using blockchain. The metadata in the form of digital signatures and timestamps is stored in decentralized ledgers. The royalty management of music artists is performed by smart contracts. The predefined royalty payment rules are encoded in the blockchain. The blockchain-based music metadata management ensures fair compensation for the artists.

Finance: In various countries, blockchain technology allows many banks and financial institutions to enable cross-border transactions promptly and with minimal cost. Blockchain-based payment protocols manage transaction metadata safely and securely. The financial products and instruments that are used by various corporations to provide international trade and commerce are secured by blockchain technology. Here, because of managing and recording the metadata using blockchain technology, processing time and paperwork are reduced.

Healthcare: In the pharmaceutical industry, the drug journey can be traced using blockchain. While tracking the drug from manufacturer to distributor and from distributor to consumer, at each stage, blockchain ensures authenticity and prevents the forging of the drugs from entering the supply chain. The patient's medical records are managed securely using a blockchain-based medical record management system. Patients can control access to their private data, thus enhancing security and privacy in the healthcare digital system.

Supply Chain: Organizations employ blockchain technology to trace the origin of items along the whole supply chain, allowing customers to confirm the validity and quality of the products. By capturing metadata including origin, processing, and shipping details, blockchain can improve food traceability and help in the quick detection of food safety concerns.

Real Estate: In real estate transactions, properties are transferred from the owner to consumer. Blockchain plays an important role in maintaining immutable and transparent property ownership records in a secure and faster way. Blockchain-based smart contracts, such as the Ethereum platform, help in managing and automating real estate rental agreements. It ensures the transparency of transactions without the need for intermediaries.

In the healthcare domain, the blood bank management system is the backbone, as it provides blood-related products for transfusion in many critical medical situations. The challenge is to ensure the availability of secure blood products and to maintain operational efficiency. The public Distribution System is a part of the government's foodgrain distribution system at affordable prices. This system plays an important role in the government's policy for the management of the food economy in the country. The following sections provide blockchain-based metadata management solutions to tackle problems in traditional systems of blood bank management in the healthcare domain and public distribution systems in the government's food policy implementation domain.

2.5.2 *Blood Bank Metadata Management Using Blockchain*

One of the most important bodily fluids is blood. It helps provide the vital nutrients needed for life to the organs. Since there is a greater need for blood than for any other medical necessity, governments frequently organize awareness campaigns to inform the public about the value of blood donation. Worldwide, blood donations total about 118.54 million. Approximately 13,300 blood centers spread across 169 nations claim to have collected 106 million donations in total. Although receiving a blood transfusion can save lives and enhance health, many individuals do not have timely access to healthy blood. Every nation's national healthcare infrastructure and policy should include the provision of safe and sufficient blood [9]. But as new diseases have emerged, there is a greater need than ever for an effective blood donation management system [10]. The Blood Management System is a big and difficult job. The efficient operation of the supply chain is restricted by the limitations and inadequacies of the current blood management system. According to the study [11], blood component waste rates typically range from 1 to 5%.

A typical blood donation flow method is shown in Fig. 2.4. First, there are three ways for donors to provide blood. The first option is via medical facilities that house blood units carried to the closest blood donation facility. In the second option, the donor donates blood at a mobile blood donation vehicle. In the third option, the donor uses blood banks directly. Every donation of blood is then processed through testing, storage, and separation in the blood bank. Using centrifuges, whole blood units are divided into their constituent parts, platelets, red blood cells, and plasma, to begin the separation process. Testing is then done to confirm blood type and identify any infectious illnesses. Units suitable for transfusion are thereafter tagged and kept in cool and freeze rooms or freezer lockers after test findings are determined [6].

The guidelines are set forth by the American Association of Blood Banks (AABB) and the Food and Drug Administration (FDA). As per these guidelines, after the separation of various components of blood, red blood cells are kept in refrigerators at 6°C with a 42-day expiration date [12]. After that, the blood units are placed into transporters and packaged by physician orders for the treatment of their patients.

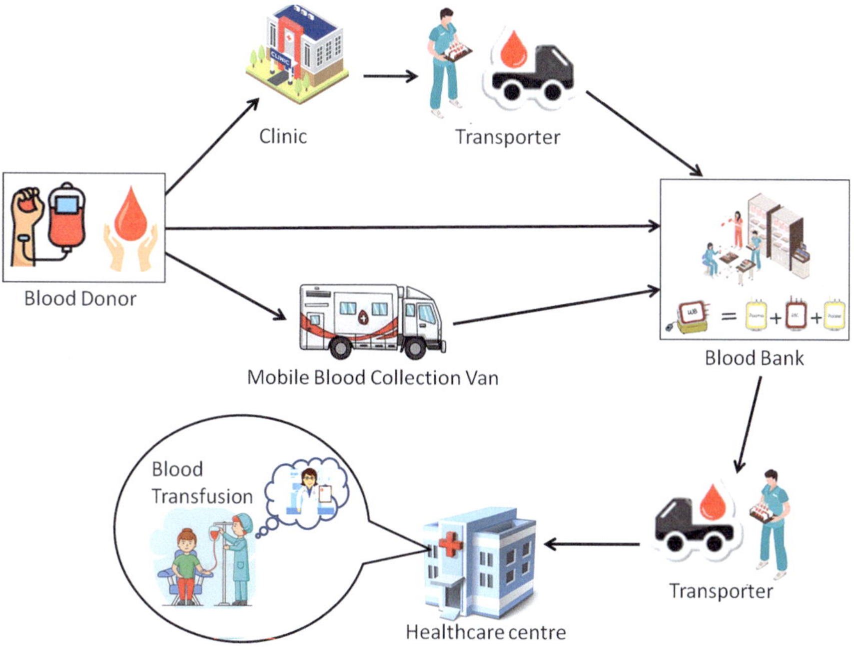

Fig. 2.4 Existing blood donation system

Finally, as per the prescription of the physician, the nurse at the hospital transfuses blood to patients after receiving blood units.

Blood-related metadata includes blood type, blood state, blood donor's medical history, blood donation date, and other relevant readings. Hospitals frequently worry about receiving incorrect blood or blood contaminated with hepatitis, HIV, or other related illnesses, even with the blood transfers done through authenticated blood banks [13]. Various risks are associated with contaminated blood being circulated through the supply chain.

The components of the blockchain-based approach for the blood donation metadata management system are shown in Fig. 2.5. The details of the components are described in Table 2.1. A blockchain hash is a piece of data or a document's digital fingerprint. It is employed to confirm whether the metadata associated with the single unit of blood has undergone any sort of alteration or tampering. A cryptographic procedure that compares an input block of data with a previously created hash value yields a blockchain hash (Fig. 2.6).

The algorithm for creating a Blockchain-based Blood Bank Metadata Management System is as follows.

Step 1: Block

- Init(Block, Index, Transactions, Timestamp, PreviousHash, Metadata) → BlockInstance

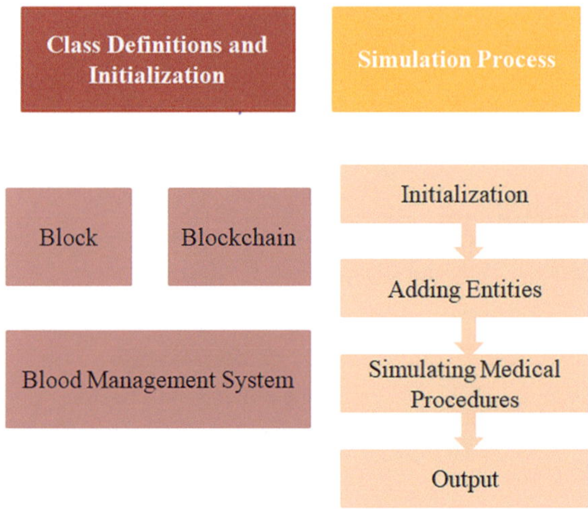

Fig. 2.5 Components of blockchain-based blood bank metadata management system

Table 2.1 Components of Blood Bank Blockchain system and its details

Component	Details of components
Block	Index, transactions, timestamp, previous hash, and metadata
Blockchain	Block instance, transactions, and metadata
Metadata of blood management system	Patient ID, name, blood type, conditions regarding blood preservation, requirement of blood, donor ID, donor name, donor's blood type, doctor ID, nurse ID, transporter ID

```
# Initialize the Blockchain and Blood Management System
blood_chain = Blockchain()
blood_management = BloodManagementSystem(blood_chain)

# Adding entities
blood_management.add_patient(1, "Radha", "A+", True)
blood_management.add_donor(11, "Sia", "A+")
blood_management.add_patient(2, "Sandeep", "B-", False)
blood_management.add_donor(12, "Nishant", "A+")
blood_management.assign_transporter(101, "Raju")
blood_management.assign_transporter(102, "Madhu")
blood_management.assign_doctor(201, "Dr. Chawla")
blood_management.assign_nurse(301, "Jenny")
blood_management.assign_nurse(302, "Saba")

# Simulate specific medical procedures
blood_management.simulate_medical_procedure(11, 1, 101, 201, 301)  # Using ID's to correctly simulate the procedure
blood_management.simulate_medical_procedure(12, 1, 102, 201, 302)  # Using ID's to correctly simulate the procedure
# Print the blockchain
print("\nBlockchain:")
pprint.pprint(blood_chain.display_chain())
```

Fig. 2.6 Simulation code snippet for blockchain-based blood bank metadata management system

- ComputeHash(BlockInstance) → Hash

Step 2: Blockchain

- Init() → BlockchainInstance
- CreateGenesisBlock(BlockchainInstance) → GenesisBlock
- AddBlock(BlockchainInstance, Transactions, Metadata) → NewBlock
- DisplayChain(BlockchainInstance) → ChainData

Step 3: BloodManagementSystem

- Init(BlockchainInstance) → BloodManagementInstance
- AddPatient(BloodManagementInstance, PatientID, Name, BloodType, NeedsBlood) → Patient
- AddDonor(BloodManagementInstance, DonorID, Name, BloodType) → Donor
- AssignTransporter(BloodManagementInstance, TransporterID, Name) → Transporter
- AssignDoctor(BloodManagementInstance, DoctorID, Name) → Doctor
- AssignNurse(BloodManagementInstance, NurseID, Name) → Nurse
- SimulateMedicalProcedure(BloodManagementInstance, DonorID, PatientID, TransporterID, DoctorID, NurseID) → MedicalProcedures

The blockchain simulation result is as follows (Fig. 2.7).

2.5.3 Public Distribution System Metadata Management Using Blockchain

One of the biggest problems the world is currently facing is food security. The availability, accessibility, consistency in the supply, and provision of basic food for all people at all times, which is also affordable for maintaining a healthy life, are the fundamental elements of food security [14]. Access to basic food for the approximately nine billion people in the world is still a major concern today [15]. India produces more food than its population consumes, but still about 200 million people suffer from undernourishment, which is around 24% of global hunger [16].

India uses the Public Distribution System (PDS) to give food grains and other necessities to the underprivileged at a discounted cost. However, PDS has a number of problems, including theft and other operational inefficiencies as well as difficulties in identifying targeted users [17]. Maintaining traceability in the supply chain, assuring effective data flow among stakeholders, and integrating information technology systems in PDS are all extremely difficult tasks for the government of India. The application of blockchain technology is also suggested by the National Institution for Transforming India (NITI Aayog) committee [18].

```
Blockchain:
[{'Current Hash': '4eda8eaf4a23ed9a8cc52a2541de8c0aa04afcb2ce57d5cf5072a8d41105f27b',
 'Index': 0,  'Metadata': {'creator': 'System', 'purpose': 'Initialize Blockchain'},
 'Previous Hash': '0',  'Timestamp': 'Sun May 12 10:36:00 2024',  'Transactions': ['Genesis Block']},
 {'Current Hash': '6c3321eba990f0c7dd2b6966df0219e8b425b428226d233a97bd6c5bf9d77829',
 'Index': 1,  'Metadata': {'event': 'Patient Registration'},
 'Previous Hash': '4eda8eaf4a23ed9a8cc52a2541de8c0aa04afcb2ce57d5cf5072a8d41105f27b',
 'Timestamp':'Sun May 12 10:36:00 2024',  'Transactions': ['Patient Added: ID=1, Name=Radha, Blood Type=A+,
Needs Blood=True']},
 {'Current Hash': 'e317dfe58aa59028678e2b6c6172b316316bd891f1f502aa52ddb02f5d7d7ce6',
 'Index': 2,  'Metadata': {'event': 'Donor Registration'},
 'Previous Hash': '6c3321eba990f0c7dd2b6966df0219e8b425b428226d233a97bd6c5bf9d77829',
 'Timestamp':'Sun May 12 10:36:00 2024',  'Transactions': ['Donor Added: ID=11, Name=Sia, Blood Type=A+']},
 {'Current Hash': '3a705db7e8e0999cdbb692b7f3d6ec2c61e31517f3bbc2f82608546150e18cdd',
 'Index': 3,  'Metadata': {'event': 'Patient Registration'},
 'Previous Hash': 'e317dfe58aa59028678e2b6c6172b316316bd891f1f502aa52ddb02f5d7d7ce6',
 'Timestamp':'Sun May 12 10:36:00 2024',  'Transactions': ['Patient Added: ID=2, Name=Sandeep, Blood Type=B-,
Needs 'Blood=False']},
 {'Current Hash': 'c37456924f7cc8e512ddd30d550c4121b2622c7b51ec1db462932e6640d9fb52',
 'Index': 4,  'Metadata': {'event': 'Donor Registration'},
 'Previous Hash': '3a705db7e8e0999cdbb692b7f3d6ec2c61e31517f3bbc2f82608546150e18cdd',
 'Timestamp':'Sun May 12 10:36:00 2024',  'Transactions': ['Donor Added: ID=12, Name=Nishant, Blood Type=A+']},
 {'Current Hash': '59c9f6fca40503b55ab4ad291e7ef7eb77a6cdfd64602e0a559c2f7032304218',
 'Index': 5,  'Metadata': {'event': 'Transporter Assignment'},
 'Previous Hash': 'c37456924f7cc8e512ddd30d550c4121b2622c7b51ec1db462932e6640d9fb52',
 'Timestamp':'Sun May 12 10:36:00 2024',  'Transactions': ['Transporter Assigned: ID=101, Name=Raju']},
 {'Current Hash': '9a9989972ee0f9de3acf0afda87c3768aabd7b213423ab6975f16530c36f2535',
 'Index': 6,  'Metadata': {'event': 'Transporter Assignment'},
 'Previous Hash': '59c9f6fca40503b55ab4ad291e7ef7eb77a6cdfd64602e0a559c2f7032304218',
 'Timestamp':'Sun May 12 10:36:00 2024',  'Transactions': ['Transporter Assigned: ID=102, Name=Madhu']},
 {'Current Hash': '36c1ba010e9cfb762e0e28cc69d80e87424e75a425d6e2873132521ad2c269f7',
 'Index': 7,  'Metadata': {'event': 'Doctor Assignment'},
 'Previous Hash': '9a9989972ee0f9de3acf0afda87c3768aabd7b213423ab6975f16530c36f2535',
 'Timestamp':'Sun May 12 10:36:00 2024',  'Transactions': ['Doctor Assigned: ID=201, Name=Dr. Chawla']},
 {'Current Hash': 'eb7e25778687bb67e3c008e20aa8818abc75a8bee13c957ffe8e48024c1ff3be',
 'Index': 8,  'Metadata': {'event': 'Nurse Assignment'},
 'Previous Hash': '36c1ba010e9cfb762e0e28cc69d80e87424e75a425d6e2873132521ad2c269f7',
 'Timestamp':'Sun May 12 10:36:00 2024', 'Transactions': ['Nurse Assigned: ID=301, Name=Jenny']},
 {'Current Hash': '0cdbcd175a439062e0eb276e8a0aadae43fc984f5d51f4c3d677e8dd9e1a9a97',
 'Index': 9,  'Metadata': {'event': 'Nurse Assignment'},
 'Previous Hash': 'eb7e25778687bb67e3c008e20aa8818abc75a8bee13c957ffe8e48024c1ff3be',
 'Timestamp':'Sun May 12 10:36:00 2024',   'Transactions': ['Nurse Assigned: ID=302, Name=Saba']},
 {'Current Hash': 'cebe4eefea4e5176a11a03676f2b5ff23dda5a383c487bdf32b6b701c4129e45',
 'Index': 10,  'Metadata': {'event': 'Detailed Medical Procedure'},
 'Previous Hash': '0cdbcd175a439062e0eb276e8a0aadae43fc984f5d51f4c3d677e8dd9e1a9a97',
 'Timestamp':'Sun May 12 10:36:00 2024',   'Transactions': ['Blood Collected from Donor ID=11 by Transporter
ID=101']},
 {'Current Hash': 'cfec77e654a9213fc03ce29d23dddca19f200135b4f4d73711c2f69c5b852a98',
 'Index': 11,  'Metadata': {'event': 'Detailed Medical Procedure'},
 'Previous Hash': 'cebe4eefea4e5176a11a03676f2b5ff23dda5a383c487bdf32b6b701c4129e45',
 'Timestamp':'Sun May 12 10:36:00 2024',   'Transactions': ['Blood Delivered to Patient ID=1 by Transporter
ID=101']},
 {'Current Hash': 'edea597aa52e84cbe322718161b79ba59a82b618d338c384b92ddd51ba6aec5b',
 'Index': 12,  'Metadata': {'event': 'Detailed Medical Procedure'},
 'Previous Hash': 'cfec77e654a9213fc03ce29d23dddca19f200135b4f4d73711c2f69c5b852a98',
 'Timestamp':'Sun May 12 10:36:00 2024',   'Transactions': ['Doctor ID=201 Prescribed Blood to Patient ID=1']},
 {'Current Hash': '563f5445e3f72d18502aa29ff3e52c638caf58eb1a15fee5c0da2025a8f4b867',
 'Index': 13,  'Metadata': {'event': 'Detailed Medical Procedure'},
 'Previous Hash': 'edea597aa52e84cbe322718161b79ba59a82b618d338c384b92ddd51ba6aec5b',
 'Timestamp':'Sun May 12 10:36:00 2024',   'Transactions': ['Nurse ID=301 Transfused Blood to Patient ID=1']}]
```


Fig. 2.7 Outcome of simulation of blood bank metadata management system

2.5.3.1 Role of Technology in PDS

There are several phases to Indian PDS, including distribution, storage, transportation, and procurement. The existing structure of Indian PDS is shown in Fig. 2.8.

The Food Corporation of India (FCI) is the designated agency that manages all procurement, transportation, storage, and distribution tasks on behalf of the Indian government. Foodgrains are purchased by FCI either directly from farmers or via buying centers or Mandis operated by state government agencies. Food grains are kept in the procuring state's FCI warehouses. Based on demand and stock from the previous period, FCI sends food grains to the warehouses of the deficit state or region in order to meet the demand of those areas of the country.

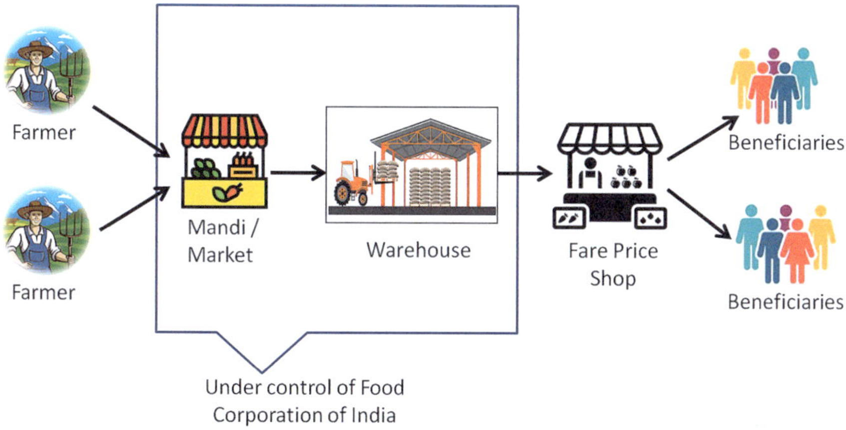

Fig. 2.8 Existing structure of Indian PDS

To serve the beneficiaries, the state government distributes the received commodities to each district and then to each Fair Price Shop (FPS), which is the retail ration shop in the state. A beneficiary pays the government-subsidized price for food grains at any nearby FPS. There have occasionally been concerns about the distribution of low-quality goods. To handle such a situation, the beneficiaries must be aware of the product's origins and quality, which is the product's whole life cycle. Till today, there is no such process set in PDS to keep track of product information from the point of origin to the FPS. Because of the erroneous beneficiary identification, the government also lacks access to trustworthy data needed to execute efficient PDS so that the nation's truly needy people will be served properly.

Throughout the supply chain of PDS, records of the transaction data are maintained manually, making it vulnerable to manipulation and human mistakes. A few problems with the existing PDS are lack of authentic identification of beneficiaries, duplicate ration cards, leakage of food grains during the process, lack of a real-time monitoring system, and lack of a decentralized database to store data [19]. The use of technology has solved these problems partially. The main technological initiatives that the Government of India has implemented with the assistance of state governments are end-to-end computerization of the supply chain, online allocation of foodgrains to FPS, the use of global positioning system (GPS) technology for the foodgrains' doorstep delivery to the FPS, and the digitization of the ration card database [20].

Along the PDS supply chain, from the point of origin to retail locations and ultimately to the customer, the suggested blockchain-based system generates data in a secure, timestamped, and immutable manner. Table 2.2 shows the details of the various components used in blockchain-based PDS.

The components of the blockchain-based PDS metadata management system are shown in Fig. 2.9.

Table 2.2 Components of PDS and its details

Component	Details of components
Block	Proof, previous hash, transactions
Blockchain	Proof-of-work, previous hash, transaction proof, transactions
Metadata of PDS	Sender, receipt, grain type, amount

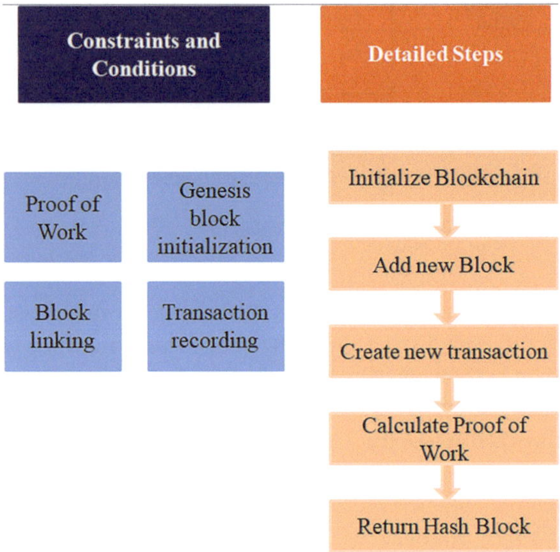

Fig. 2.9 Components of blockchain-based PDS metadata management system

The stepwise algorithm for a blockchain-based PDS metadata management system is as follows (Figs. 2.10 and 2.11).

1. **Initialize Blockchain**
   - BC = init_blockchain()
   - This function initializes the blockchain with a genesis block.
2. **Add New Block**

   Inputs: $Proof, PreviousHash, Transactions$
   B=new_block($Proof, PreviousHash, Transactions$)
   This function creates and returns a new block using the provided proof, previous hash, and the list of transactions.
3. **Create New Transaction**

   Inputs: $Sender, Recipient, GrainType, Amount$
   TID=new_transaction($Sender, Recipient, GrainType, Amount$)
   This function records a new transaction and returns a transaction ID, which indicates the next block index to hold this transaction.

```
# Initialize blockchain
blockchain = Blockchain()

# Transactions and blocks for Sakharam, FCI, FPS, Ramu, and Sadashiv
blockchain.new_transaction("Sakharam", "FCI", "Wheat", 1000)
blockchain.new_transaction("Sakharam", "FCI", "Jowar", 600)
proof = blockchain.proof_of_work(blockchain.last_block['proof'])
previous_hash = blockchain.hash(blockchain.last_block)
blockchain.new_block(proof, previous_hash)

blockchain.new_transaction("FCI", "FPS A", "Wheat", 500)
blockchain.new_transaction("FCI", "FPS A", "Jowar", 300)
blockchain.new_transaction("FCI", "FPS B", "Wheat", 500)
blockchain.new_transaction("FCI", "FPS B", "Jowar", 300)
proof = blockchain.proof_of_work(blockchain.last_block['proof'])
previous_hash = blockchain.hash(blockchain.last_block)
blockchain.new_block(proof, previous_hash)

blockchain.new_transaction("FPS A", "Ramu", "Wheat", 50)
blockchain.new_transaction("FPS A", "Ramu", "Jowar", 10)
blockchain.new_transaction("FPS B", "Sadashiv", "Jowar", 40)
proof = blockchain.proof_of_work(blockchain.last_block['proof'])
previous_hash = blockchain.hash(blockchain.last_block)
blockchain.new_block(proof, previous_hash)
```

Fig. 2.10 Simulation code snippet for blockchain-based PDS metadata management system

4. **Calculate Proof-of-Work**
   - Input: $LastProof$
   - $Proof$ = proof_of_work($LastProof$)
   - This function computes the proof required for creating a new block based on the proof of the last block.
5. **Hash Block**
   - Input: $Block$Block
   - H = hash($Block$)
   - This function computes and returns the hash of the block.

```
Current Hash: c5ebc2f46c2098e5b10cf4d98bde1d35925e475b1a5b68b6057dc3424d46f5e4, Block Index:
1,Timestamp: 1715518850.569415,Transactions: []
Previous Hash: 1

Current Hash: ccba161b4ab3d8e23bc076187bbad5a37b7cdb4583c041e2a94d90dd1d3afa1a, Block Index:
2,Timestamp: 1715518850.6269639,Transactions: [
   {"sender": "Sakharam“, "recipient": "FCI“, "grain_type": "Wheat“, "amount": 1000},
   {"sender": "Sakharam“, "recipient": "FCI“, "grain_type": "Jowar“, "amount": 600 }]
Previous Hash: c5ebc2f46c2098e5b10cf4d98bde1d35925e475b1a5b68b6057dc3424d46f5e4

Current Hash: aa4a90d854aa19cd1bfac59216334bd1d576462e6c713d0cf792aa9ec2a0c03f, Block Index:
3,Timestamp: 1715518850.6803315,Transactions: [
   {"sender": "FCI“, "recipient": "FPS A“, "grain_type": "Wheat“, "amount": 500},
   {"sender": "FCI“, "recipient": "FPS A“, "grain_type": "Jowar“, "amount": 300 },
   {"sender": "FCI“, "recipient": "FPS B“, "grain_type": "Wheat“, "amount": 500 },
   {"sender": "FCI“,  "recipient": "FPS B“, "grain_type": "Jowar“, "amount": 300  }]
Previous Hash: ccba161b4ab3d8e23bc076187bbad5a37b7cdb4583c041e2a94d90dd1d3afa1a

Current Hash: 129b1cf60a2ffc34863093d1f0a3221b1b6369606633e9c2e15c7abdd0ee6bef, Block Index:
4,Timestamp: 1715518850.8831723,Transactions: [
   {  "sender": "FPS A“, "recipient": "Ramu“,  "grain_type": "Wheat“, "amount": 50  },
   {  "sender": "FPS A", "recipient": "Ramu",  "grain_type": "Jowar", "amount": 10  },
   {  "sender": "FPS B“, "recipient": "Sadashiv“, "grain_type": "Jowar", "amount": 40  }]
Previous Hash: aa4a90d854aa19cd1bfac59216334bd1d576462e6c713d0cf792aa9ec2a0c03f

{'index': 4, 'timestamp': 1715518850.8831723,
'transactions': [{'sender': 'FPS A‘,   'recipient': 'Ramu‘,   'grain_type': 'Wheat‘,   'amount': 50},
  {'sender': 'FPS A‘,   'recipient': 'Ramu‘,   'grain_type': 'Jowar‘,   'amount': 10},
  {'sender': 'FPS B‘,   'recipient': 'Sadashiv‘,   'grain_type': 'Jowar‘,   'amount': 40}],
 'proof': 119678,
 'previous_hash': 'aa4a90d854aa19cd1bfac59216334bd1d576462e6c713d0cf792aa9ec2a0c03f'}
```

Fig. 2.11 Outcome of simulation of PDS metadata management system

2.6 The Future of Metadata Management with Blockchain

2.6.1 Emerging Trends and Future Potentials of Blockchain in Metadata Management.

Decentralized Metadata System: Blockchain's decentralized architecture allows the data and metadata to be kept across the network by eliminating the dependency on a central authority. As the modification to the metadata requires consensus from all the network users, the centralization enhances the integrity of the data. Self-sovereign identity (SSI) solutions and decentralized identification systems (DIDs) are two examples.

Interoperability: It becomes important as the number of blockchains increases. The cross-chain bridges and protocols help to smooth the transfer of data between various blockchains. The protocols used are Polkadots and Cosmos. Interoperability refers to the sharing and accessing of metadata among blockchain networks in metadata management.

Verifiable credentials: Blockchain technology can issue and verify digital credentials such as licenses, degrees, and professional certifications. These authenticated login credentials are tamper-proof and encrypted. Credential-related metadata, such as issuer, the recipient, and expiration date, can be kept in check.

Smart contract for the governance of metadata: The processes of metadata management are automated through smart contracts. For example, a smart contract can guarantee privacy standards are followed by applying access control rights for metadata. The management of metadata becomes efficient, transparent, and auditable.

Techniques for privacy enhancement: Zero-Knowledge Proofs (ZKPs) enable metadata validation without disclosing the actual data. Privacy-focused blockchains like Monero and ZCash preserve transparency while protecting the metadata. Effective metadata management requires a balance between privacy and transparency.

The evolving landscape of data management and how blockchain fits in

The landscape of data management is continuously evolving with changes in technology as well as regulatory compliance. Due to the unique properties to fit into the dynamic environment blockchain technology is preferred.

The unique properties of blockchain to fit into the evolution of data management are as follows.

Decentralization: Conventional data management system depends on centralized databases, which causes single-point failure and also increases vulnerability in the system, whereas the blockchain distributes the data throughout the network. Hence, it helps to resist attacks and failures. With the decentralized data management in blockchain, stakeholders from diverse organizations or sectors can share data securely without the need for any centralized system. It facilitates streamlining processes such as financial transactions, health records sharing, and supply chain management by ensuring enhanced data-sharing capabilities.

Immutability and integrity: In the blockchain, once the information is recorded for modification, it requires the agreement of the entire network by providing a trustworthy audit trail. The change in the state of the records has a great deal of significance for the sectors where trust is the highest preference, such as financial services, legal services, and healthcare services. The Consistency and Reliability of the data over time are guaranteed by the immutability of the blockchain. It reduces the data decay and, hence, makes the analytical insights and reporting more accurate.

Transparency and traceability: The famous transparent ledger, which is the blockchain, is the main reason for the durability of the blockchain through the data transactions that are verified and recorded in a detailed monitoring and auditing system. This feature is very essential for the pharmaceutical industry, where the product journey has to be tracked in the most detailed way because of compliance and safety issues. The blockchain is a technology that gives proof of the integrity of the consent trails and data access logs, thus ensuring transparency and trust. Hence, the regulatory compliance that is critical for data protection is obtained.

Automation via smart contract: Smart contracts can ease the work of data management, such as record updates, transaction execution, and compliance enforcement. Therefore, it enhances the effectiveness and the precision of the processes. Besides, the amount of manual monitoring that is required can be reduced by automating these tasks, which, in turn, leads to cost reduction, and the overall expenses with data management are lowered.

Privacy and Security: Blockchain can combine with the most advanced encryption techniques to enhance data security and privacy. The personal, private, and business data management tasks turn out to be the most significant ones when you are at work. Blockchain allows the users to be their own masters, i.e., what type of data they want to share and with whom, hence, they are the ones to decide. The rules are made for that reason, and the one who enforces them does it because he/she is personally interested in protecting his/her own safety.

Interoperability: The organizations are going to build a blockchain system that will be interoperable; thus, the data will be exchanged among different platforms and industries, and the data management will be more efficient and easier.

The blockchain becomes the main piece of the data management of the future. The device finds the solutions to the problems, which are usually issues with the regular system. Technology is on the way, and at the same time, it is necessary to have data management strategies. The numerous projects that will be initiated the soonest will be based on the following possible developments.

Boosting Scalability: The improvement of blockchain solutions, like sharding, layer 2 protocol, and consensus mechanisms, can be done by using them to solve the existing problems of the scalability of blockchain.

Regulatory Progress: The increase in the use of blockchain is leading, and, therefore, the possibility of the emergence of special regulatory standards for the management of blockchain data will be high. These principles will be the key to resolving the legal issues; thus, they will be the main factor in the huge tech growth.

Hybrid Systems: The conventional databases and the blockchain are being amalgamated to provide a system that will be the data, and thus, the hybrid data management system will be created. This system is a combination of both methodologies, and it, in turn, gets the strengths that both methodologies have, thus it carries the advantages that they both have.

2.7 Conclusion

Using blockchain technology, the metadata management industry is transforming, where data security, transparency, and efficiency are handled in an improved manner. Using metadata management, blockchain technology is becoming more versatile and transparent. Several benefits of metadata management have been witnessed. Creating a scalable and efficient blockchain system that can handle large datasets and also provide high throughput in a distributed environment can be possible with the help of an efficient consensus mechanism. By using the specialized regulatory framework, blockchain technology can be embedded into a vital infrastructure. This is going to help solve the metadata management use cases and promote wider adoption of it. Using blockchain technology in blood management systems helps to provide critical metadata such as donor records and blood transfusion tracking. Similarly, in public distribution systems, blockchain can be used to reduce fraud and to ensure appropriate resource distribution. The blockchain system acquires stakeholder's trust because of its characteristics such as decentralized structure, immutable records, an auditable framework, and managing sensitive and critical data about the system.

To avail of the benefits of traditional database and blockchain technology, hybrid data management systems are suggested. Due to the blockchain capability, the trust it has gained helps the wider adoption of blockchain across industries like supply chain, public administration, education, healthcare, and finance. To enhance the potential of blockchain and metadata management, technologies like AI and IoT can be used. This will create a dynamic, efficient, responsive, and intelligent system in the digital world to unlock the power of computation.

References

1. Habermann, T.: Metadata life cycles, use cases, and hierarchies. Geosciences **8**(5), 179 (2018)
2. Kemp, J.: Metadata and discoverability: a use case overview. Inf. Serv. Use **38**(1–2), 81–84 (2018)
3. Krug, R.M., Petchey, O.L.: Metadata made easy: develop and use domain-specific metadata schemes by following the dmdscheme approach. Ecol. Evol. **11**(14), 9174–9181 (2021)
4. Tundjungsari, V.: Blockchain-based music metadata copyright protection using fuzzy logic. Int. J. Appl. Eng. Technol. **4**(2), 101–105 (2022)
5. Lee, S.: Pidgin metadata framework as a mediator for metadata interoperability. Libri **71**(4), 375–387 (2021)
6. Hawashin, D., Mahboobeh, D.A.J., Salah, K., Jayaraman, R., Yaqoob, I., Debe, M., Ellahham, S.: Blockchain-based management of blood donation. IEEE Access **9**, 163016–163032 (2021)
7. Peltoniemi, T., Ihalainen, J.: Evaluating blockchain for the governance of the plasma derivatives supply chain: how distributed ledger technology can mitigate plasma supply chain risks. In: Blockchain in Healthcare Today (2019)
8. Gupta, A.: Top five metadata management best practices. Collibra (2022). https://www.collibra.com/us/en/blog/metadata-management-best-practices. Accessed 20 Jan 2024
9. World Health Organization (WHO): Blood Safety and Availability (2023). https://www.who.int/news-room/fact-sheets/detail/blood-safety-and-availability#:~:text=About%20118.54%20million%20blood%20donations,vary%20according%20to%20income%20group
10. Kim, S., Kim, J., Kim, D.: Implementation of a blood cold chain system using blockchain technology. Appl. Sci. **10**(9), 3330 (2020)
11. Hannon, T.: Waste not, want not. Am. J. Clin. Pathol. **143**(3), 318–319 (2015)
12. Eligibility Requirements: (n.d.). https://www.redcrossblood.org/donate-blood/how-to-donate/eligibility-requirements.html
13. Lakshminarayanan, S., Kumar, P.N., Dhanya, N.M.: Implementation of blockchain-based blood donation framework. In: International Conference on Computational Intelligence in Data Science, pp. 276–290. Springer International Publishing, Cham (2020)

14. Parameswari, C.D., Ilayaraja, M.: A secure and intelligent public distribution system (SIPDS) based on deep learning and Ethereum using predictive analytics for supply chain services. Soft Comput. 1–17 (2023)
15. Singh, S.K., Jenamani, M., Dasgupta, D., Das, S.: A conceptual model for Indian public distribution system using consortium blockchain with on-chain and off-chain trusted data. Inf. Technol. Dev. **27**(3), 499–523 (2021)
16. Bradford, N., Wolfe, D.A.: Governing regional economic development: innovation challenges and policy learning in Canada. Camb. J. Reg. Econ. Soc. **6**(2), 331–347 (2013). https://doi.org/10.1093/cjres/rst006
17. Niehaus, P., Atanassova, A., Bertrand, M., Mullainathan, S.: Targeting with agents. Am. Econ. J. Econ. Pol. **5**(1), 206–238 (2013)
18. NITI Aayog: National Strategy for Artificial Intelligence, pp. 71–80 (2018). http://niti.gov.in/writereaddata/files/document_publication/NationalStrategy-for-AI-Discussion-Paper.pdf
19. Biswal, A.K., Jenamani, M.: Leveraging ICT for food security: an analysis in the context of PDS in India. In: Social Transformation–Digital Way: 52nd Annual Convention of the Computer Society of India, CSI 2017, Kolkata, India, January 19–21, 2018, Revised Selected Papers 52, pp. 376–390. Springer Singapore (2018)
20. Bhattacharya, S., Falcao, V.L., Puri, R.: The public distribution system in India. In: The 1.5 Billion People Question, vol. 43 (2017)
21. Le, T.H., Trong, P.N., Gia, K.H., Vo, H.K., Huong, L.H., Dang, K.T., et al.: BloodMan-chain: a management of blood and its products transportation based on blockchain approach. In: International Conference on Parallel and Distributed Computing: Applications and Technologies, pp. 199–209). Springer Nature Switzerland, Cham (2022)
22. Le, H.T., Nguyen, T.T.L., Nguyen, T.A., Ha, X.S., Duong-Trung, N.: Bloodchain: a blood donation network managed by blockchain technologies. Network **2**(1), 21–35 (2022)
23. Sandaruwan, P.A.J., Dolapihilla, U.D.L., Karunathilaka, D.W.N.R., Wijayaweera, W.A.D.T.L., Rankothge, W.H., Gamage, N.D.U.: Towards an efficient and secure blood bank management system. In: 2020 IEEE 8th R10 Humanitarian Technology Conference (R10-HTC), pp. 1–6. IEEE (2020)
24. George, N.A., McKay, F.H.: The public distribution system and food security in India. Int. J. Environ. Res. Public Health **16**(17), 3221 (2019)

Chapter 3
Blockchain and IOT Integration: Prospects and Obstacles

M. Banurekha

Abstract The Internet of Things (IoT) is a network of physical objects that are linked together over the internet in order to gather data and make deft choices. Nevertheless, IoT is vulnerable due to a lack of internal safeguards, and its overuse has resulted in a number of security and privacy concerns. Blockchain (BC), with its "security by design," can assist in meeting important security needs within Internet of Things. Additionally, decentralized, blockchain ledgers are kept up to date at every network node. Because of its flexibility and security, blockchain technology makes managing large systems on it easier while still providing the benefits of decentralization. The majority of IoT's architectural flaws may be fixed with the aid of BC features, including immutability, transparency, auditability, data encryption, and operational resilience. As technology develops, this vision is coming to pass, but there are still issues that need to be fixed. This is particularly true in the security space, where data trust is concerned, and it is crucial to make sure that this excellent supply of data materializes, given the anticipated growth of the Internet of Things in the upcoming years. In this article, we look into a few IoT operational frameworks, whether or not they can be integrated with blockchain technology, and precautions staff need to take while deploying and using IoT and BCN. We also discuss underlying security issues and suggestions for IoT networks with blockchain integration.

Keywords Blockchain (BC) · Internet of things (IOT) · Cryptography · Security · Data encryption · Blockchain ledgers

3.1 Introduction

The financial sector has recently been impacted by the swift development of digital currencies and blockchain technology, which has given rise to a new crypto economy. The internet, finance, business, and medical center security are just a few of the sectors where these applications are expanding. Furthermore, the use of IoT [1] has

---

M. Banurekha (✉)
Dr.N.G.P.Arts and Science College, Coimbatore, Tamilnadu, India
e-mail: banurekha@drngpasc.ac.in

P. M. Jeyanthi (ed.), *Distributed Ledger Technology in Communication*, Signals and Communication Technology, https://doi.org/10.1007/978-3-032-07682-3_3

increased globally due to the growth of urban development. The Internet of Things (IoT) is now a collection of technologies that include radio frequency identification (RFID), wireless sensor networks (WSN), and other tools for communication, identification, and exploitation online. These days, wearables, electronic gadgets, and hardware development platforms can all be considered IoT devices. The Internet of Things (IoT) is a key factor in the development of smart cities, which have several uses across all societal sectors. By 2020, 20–50 billion gadgets will be connected, according to a number of research [2] publications. This is mostly because of how many devices IoT can deploy. A completely linked world where quantifiable data may interact and communicate with items is what the Internet of Things (IoT) envisions. This makes it possible to depict the actual world digitally and opens up a wide range of smart applications across several industries, such as wearables, smart cities, healthcare, automotive, environment, smart water, and smart grid. It's evident that IoT applications in the digital sector can boost productivity since producing a lot of data requires a strong connection and energy. There are several difficulties due to limited power supply, networking, computing, and storage capacities. The rapid proliferation of Internet of Things requires standard procedures and protocols. As a result of this division, there are fewer vertical silos now, and IoT is being used to lessen the region's variety. When working with data from financial and governmental entities, it's crucial to take data trustworthiness into account in addition to the diversity and inclusion issues that the Internet of Things presents.

However, how can we guarantee that the data obtained from outside sources and other outside organizations? IoT businesses won't be misrepresented, changed, or faked in any manner? In a centralized architecture, this question is hard to address [3, 40]. Organizations that lack credibility have the ability to alter facts to suit their own needs. As a result, the data they offer can be wholly speculative. Because of this, it is necessary to confirm that the data has never been modified. Using a distributed service that all parties can rely on to guarantee data consistency is one method to assure the integrity of IoT data. There is a method to confirm that the data hasn't been altered since it was initially specified if all participants have data.

Furthermore, governments will be able to safely exchange information with individuals through the use of a system that ensures the accuracy of the data. Cutting-edge decentralized programs without the involvement of an assumed outsider have thus been made possible by the advent of smart contracts, which are personal computer conventions intended to work with, confirm, and authorize naturally the exchange and arrangement among distinct deceitful groups. Notwithstanding the benefits of intelligent agreements, several issues like security threats, defects, and legitimate worries, nevertheless undermine their acceptability. The blockchain has been described as an innovation that records every trade that has taken place in a distributed organization through a circulating data set for more than a decade. Perceived as a widely held belief system, it successfully overcomes the problem associated with the confidence of an integrated entity.

As a result, several hubs collaborate with one another in a blockchain network to get and maintain a number of shared trade records in a distributed fashion without relying on a third party. The first suggested digital currency, Bitcoin, was introduced by Satoshi Nakamoto in 2008, portraying the blockchain as a relevant technological advancement. It allowed users to transfer bitcoins—cryptographic money—securely and independently of a central authority. Furthermore, blockchain-based foundations for the digital currency, Ethereum, NXT, and Hyperledger Fabric—the largest open source blockchain in the world, created by the Linux Establishment, a contributor that included companies like IBM—were also suggested [4]. Unlike Bitcoin, they may make use of cunning arrangements. The terms of agreements between at least two parties are covered by blockchain innovation, which surpasses traditional agreements due to clever agreements that automate the implementation of agreements in a dispersed environment when certain criteria are satisfied. Evensmart contracts have gained popularity lately, they still have a lot of obstacles to overcome. For instance, the Decentralized Independent Association (DAO) agreement was manipulated to extract around 2 million Ether (50 million USD at the time) because of its re-entrancy issue [2, 5]. Brilliant agreements have a variety of difficulties in addition to the weakness issue, including protection, legality, and implementation issues.

3.1.1 Principles of Blockchain

A distributed ledger system, often known as a shared ledger, in blockchain. Distributed ledger technology (DLT) is a system that shares and replicates consensus among several network nodes [21]. Each node in the network functions as a peer. When this P2P network shares a ledger database, every node copies the ledger and stores a duplicate copy, enabling the node to update itself autonomously. Because there is no central authority in this system, it is entirely decentralized. The blockchain's distinct structure, which consists of arranged blocks of grouped data, sets it apart from other distributed ledgers. Cryptographic primitives are utilized to establish a secure connection between these blocks [22]. Both a public and a private key are included in every block. All messages transmitted by the sender node (user) to all other nodes in the network are encrypted using the public key. Using a private key, the receiving nodes decode incoming messages. A suitable private key is necessary in order to decode a specific message that has been encrypted using a matching public key.

A node would broadcast the block to every other node in the network after signing (hashing) it in case of Bitcoin. Only blocks containing a predetermined sequence of zeros would be permitted to join the network. An additional bit of information called the block's nonce is added to each block in order to identify the qualifying signature—the one with a string of zeros. The receiving nodes, also known as miners, will use their computing capacity to repeatedly change the nonce in an attempt to obtain an eligible signature for the blocks that have been broadcast. The issue will be resolved as soon as the ideal nonce combination is discovered. Every other node in the network receives this right combination to verify. All nodes would correctly verify the block (whose eligible signature was just discovered) as genuine if it included a valid transaction and the hash of its prior block. Ultimately, the block would update the blockchain by being added to their (nodes) chains. It is easily seen that while new blocks of data may be added to this structure, it is not feasible to remove or change blocks of data that have already been stored.

As a result, the blockchain technology works best for keeping records, voting, tracking assets, recording events, and processing transactions. In a word, the blockchain is a system that shares distributed ledgers among many nodes in a P2P system. It consists of a series of blocks with time-stamped transactions that are validated by every node in the network and secured by public key cryptography. The blockchain is unchangeable as new elements (blocks) can be added to it at any time, but older blocks cannot be removed or changed. Blockchain technology's primary characteristics are immutability, autonomy, decentralization, transparency, and anonymity.

Blockchain's Operation

1. A new transaction gets added to the blockchain. Using both public and private keys, all necessary data is double-encrypted before being sent.
2. After that, the transaction is sent to the global network of peer-to-peer computers for verification. Every node connected to the network will verify that the transaction is legitimate and that there is enough money available to complete it.
3. Many nodes and numerous transactions are validated simultaneously in a typical blockchain network. The transaction will be added to the mempool as soon as it has been confirmed and deemed valid. A mempool is made up of all the validated transactions at a specific node, while a block is made up of many mempools.
4. In order to make a block permanent, the nodes that compose it will attempt to add it to the blockchain network. However, the operation of the blockchain network will be interfered with if all nodes are permitted to contribute blocks in this way. To address this issue, the nodes employ a consensus method to guarantee that each new block added to the blockchain represents the single, accepted version of the truth that is accepted by every node, and that only valid blocks are safely appended to the chain. We refer to the node that is chosen to add a block to the blockchain as a "miner" since they will be rewarded. In order to add a block to the blockchain, the consensus algorithm generates a hash code for it.
5. The freshly formed block is now prepared for addition blockchain once it has been authenticated and assigned a hash value. Every block includes the hash value of the one before it, which is how a blockchain is formed by cryptographically connecting all of the blocks together. The open end of the blockchain receives a fresh block.
6. The transaction is finished as soon as the block is put on the blockchain, and the information about it is kept there forever. Anybody may obtain the transaction's information and validate it (Fig. 3.1).

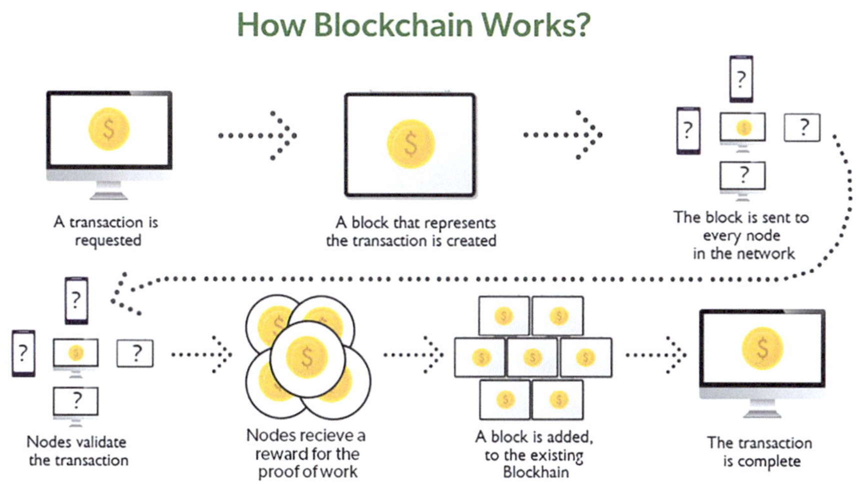

Fig. 3.1 Block chain operation

3.2 Types of Blockchain

3.2.1 Public Blockchain

1. Anyone may join and perform transactions on this distributed ledger, which operates without authorization.
2. Every peer has a copy of this version of the ledger, which is non-restrictive. This implies that it is accessible to everybody with an internet connection.
3. This user may execute mining operations and has access to both history and current records.
4. To validate transactions and add them to the ledger, these intricate calculations need to be completed.
5. No legitimate record or transaction in the blockchain network may be changed. Since the source code is often available, anybody may examine the transactions, identify issues, and recommend solutions (Fig. 3.2).

3.2.2 Applications of Public Blockchain

Voting: To ensure transparency and confidence, governments can conduct voting on a public blockchain.

Logistics Optimization: Companies use the technology to keep an eye on shipments in real time, utilizing IoT devices to keep track of things like temperature and humidity. This helps cut down on waste and make things run more smoothly.

Fig. 3.2 Public block chain

3.2.3 *Private Blockchain*

1. A blockchain network is managed by a single person or functions in a private setting, like a limited network.
2. A public blockchain network's decentralization and peer-to-peer connectivity are shared by this blockchain, although it is much smaller.
3. Instead of being accessible to anybody wishing to donate computing power, they are frequently operated on a limited network within a company or group.
4. Two other terminology for them are permissioned blockchains and enterprise blockchains (Fig. 3.3).

Fig. 3.3 Private block chain

3.2.4 Applications of Private Blockchain

Supply Chain Management: An organization's supply chain may be overseen by the use of a private blockchain. Asset Ownership: Asset tracking and verification may be done via a private blockchain. Internal Voting: It also makes internal voting feasible.

3.2.5 Hybrid Blockchain

1. Hybrid blockchains, which mix the advantages of public and private blockchains, are used by organizations that need the best of both worlds.
2. Businesses may use it to build a permission-based private system in addition to a permissionless public system, giving them control over which data is made public and to whom.
3. Transactions and records on a hybrid blockchain are normally private, but they may be verified upon request by providing access through a smart contract (Fig. 3.4).

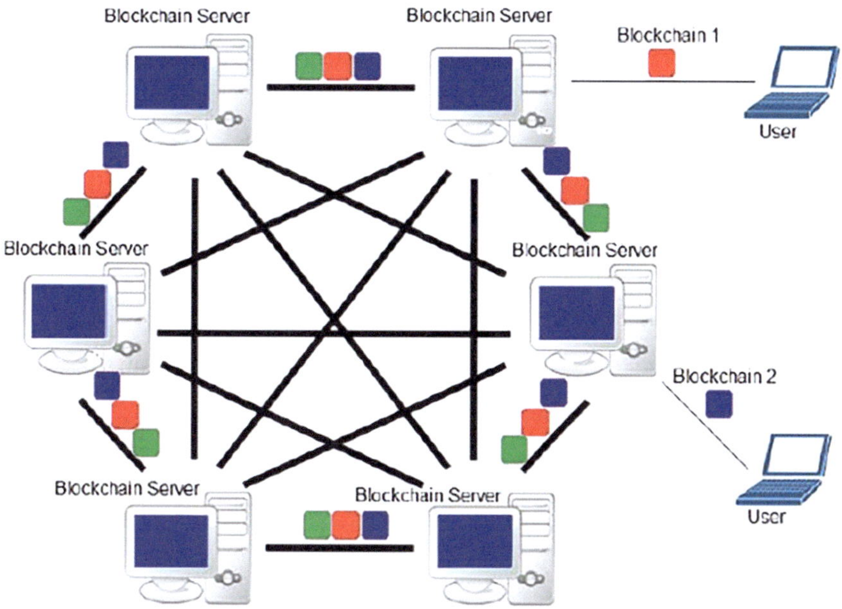

Fig. 3.4 Hybrid block chain

3.2.6 Applications of Hybrid Blockchain

Real Estate: Real estate firms may manage their systems and provide public information via hybrid networks.

Retail: Retailers can also benefit from process streamlining thanks to the hybrid network.

Highly Regulated Markets: The banking industry is one highly regulated sector where hybrid blockchains work particularly effectively.

3.2.7 Consortium Blockchain

1. A Consortium blockchain, often referred to as a federated blockchain, contains both private and public blockchain capabilities, just as a hybrid blockchain.
2. It is distinct, though, in that it entails the collaboration of several organizational members via a decentralized network.
3. In a consortium blockchain, the consensus techniques are managed by predetermined nodes.
4. Transactions are initiated, received, and validated by its validator node. Member nodes have the ability to send and receive transactions (Fig. 3.5).

3.2.8 Applications of Consortium Blockchain

Payments and Banking: Banks cooperating with one another can create a consortium. Which nodes will validate transactions is within their control.

Research: Data and findings from studies may be shared via a consortium blockchain.

Food Tracking: You may use it for that as well.

3.2.9 Consensus Mechanism

A consensus mechanism is a system that verifies a transaction and designates it as genuine. To increase traders' trust in a coin, this system maintains a list of all legitimate transactions for that coin on a blockchain. For security reasons, a number of currencies, including Bitcoin, Ethereum, and others, adopt this mechanism. It accomplishes the majority of users on a single network's agreement. The blockchain's

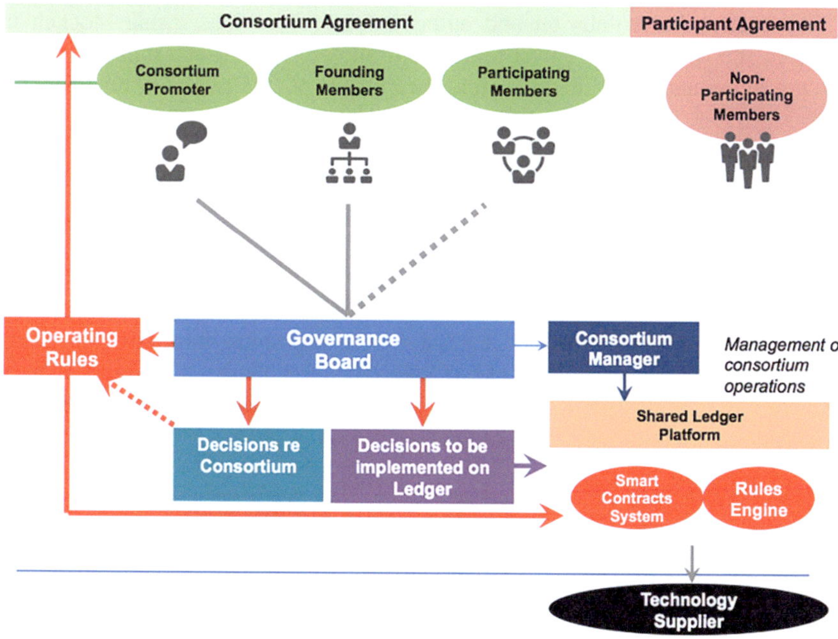

Fig. 3.5 Consortium block chain

security is preserved via the consensus process, which records all authorized transactions. Due to the decentralized nature of cryptocurrency trading, it is crucial to prevent sellers from purposefully deceiving buyers.

The consensus process makes sure that a transaction is reflected in the blockchain as soon as it is validated, which helps to increase confidence in a blockchain. In order to establish consensus inside a blockchain network, guarantee security and trust, a number of approaches are necessary. Consensus procedures also guarantee that every transaction pertaining to a currency is accurately recorded in the blockchain.

When exchanging coins, a number of processes are employed as a consensus mechanism. The following are these mechanisms:

3.2.10 Evidence of Work

"Work" refers to the steps involved in solving a very complicated problem, and "proof" refers to the answer itself. Cryptocurrency miners compete to find a solution and earn the authority to handle the transaction. The traders of these coins pay a mining fee to the fastest solver.

This system keeps tabs on and authenticates transactions inside blockchain networks. It is quite safe and gives miners the ability to validate new transactions. Its high power demand and difficulties for lone miners are only two of its drawbacks.

3.2.11 Evidence of Stake

To validate a transaction, this system selects a maximum coin owner at random. The owner is also able to make a block for the same currency. Comparatively speaking, this technique uses less energy and needs a shorter transaction time and charge. This technique is widely used by coins like as Etherium 2.0, Polkadot, Cosmos, Cardano, ThorChain, Nxt, and Algorand. There is a security concern since the sole ownership of a currency's network will pass to the owner who possesses 51% or more of the coin.

3.2.12 Evidence of Capability

The PoC process is highly dependent on the hard drive's available free space. This is due to the fact that a trader must keep a large number of solutions to a coin's hash issue. When compared to PoW and PoC methods, it is extremely efficient. These methods are used by coins like Chia, Burst, Storj, and SpaceMint.

3.2.13 Evidence of Activity

This technique combines the concepts of Proof of Stake and Proof of Work. The greatest elements of PoW and PoS have been combined in its design. The Proof-of-Activity technique works similarly to Proof of Work at first. A fresh block begins to operate similarly to a Proof-Of-Stake process once it is finished. This method is used by coins like DECR (Decred).

3.2.14 Evidence of Authority

This special method was developed by many organizations and commercial businesses. The generation of new blocks and transactions are authorized by validators who own approved accounts. To be able to authenticate a transaction, these validators have to reveal who they really are.

3.2.15 Evidence of Burn

PoB seeks to enhance blockchain's quality so that it may be widely and readily utilized as a tool for safer, quicker transactions. PoB is intended to stop fraudulent activity on a blockchain network, after PoW and PoS. This technology is used by cryptocurrencies like Bitcoin to provide traders with safe transactions.

3.2.16 Evidence of Passed Time

This system was developed by Intel Corporation that allow blockchain to determine who will build the next block. To choose who will create the next block, a lottery is used. As a result, it offers every trader an equal opportunity to start the next block. Utilizing fewer materials and consuming less energy makes it an efficient operation and safe transactions for merchants.

3.2.17 Exploring IOT (Internet of Things)

A huge and complex web that connects an inconceivably large number of devices, items, and systems is what the Internet of Things (IoT) is best described as. It's a network that reaches well beyond desktop computers, into our everyday existence, and is changing the way we engage with technology. The Internet of Things (IoT) is the network of sensors, actuators, and other technologies that connect physical items to the digital world. The increased connection and security that come with this interconnectivity are unparalleled. It is useful to think about how the IoT affects many facets of our lives in order to have a better understanding of its ramifications. First, IOT facilitates smooth data transmission between devices, enabling them to transmit information automatically and without the need for human involvement.

Motion sensors can identify patterns of occupancy, and smart home equipment such as lighting systems or thermostats can interact with one another to optimize energy use. Second, it improves surveillance capabilities in terms of safety and security by incorporating cameras and sensors into common place like traffic signs and streetlights. These networked gadgets have real-time anomaly detection capabilities and can notify authorities if needed. Finally, from an economic standpoint, the Internet of Things offers new business models that make use of device data to offer tailored services or predictive analytics. The potential of the Internet of Things is further demonstrated by drawing a representation of this networked world:

1. **Logistics and Supply Chain**: Guaranteeing Efficiency and Transparency Complex networks comprising a variety of players, including distributors, retailers, and manufacturers, define the supply chain sector. Through real-time tracking of commodities and the recording of their route from point of origin to

point of destination, blockchain-IoT integration may usher in a new era of transparency. Smart sensors can keep an eye on variables like humidity and temperature to guarantee the genuineness and quality of goods. The blockchain can initiate warnings and steps to stop possible losses in the event of any discrepancy or divergence from predetermined parameters.

2. **Medical: Strengthening Patient-Centred Care** IoT and blockchain together have the potential to completely transform patient care in the healthcare industry. Patient data may be gathered by wearable sensors and devices, and then safely stored on a blockchain. Patients have complete control over who may access their information, which improves consent handling and privacy. Additionally, by automating billing and streamlining the processing of insurance claims and smart contracts may save administrative expenses.
3. **Energy Management**: Making the Best Use of Available Resources between blockchain and IoT can help the energy sector by facilitating effective energy management. Blockchain guarantees the confidentiality and accuracy of data collected in real-time on energy use via IoT-enabled smart meters. Peer-to-peer energy trading, made possible by smart contracts, enables people to sell extra energy to other people directly, encouraging the use of renewable energy sources and the decentralization of power delivery.
4. **Agriculture**: Providing Traceability and Food Safety in the agriculture industry may be revolutionized by integrating blockchain and IoT to ensure product traceability across the supply chain. This creative method makes use of sensors to keep an eye on important variables like temperature and soil conditions, and it meticulously documents every stage of production and delivery through the use of blockchain technology. Blockchain in agriculture helps quickly identify the source of contamination in the case of a foodborne illness epidemic, hence reducing the harm to the public's health.
5. **Smart Cities**: Facilitating Ecological Urban Growth The combination of IoT with blockchain technology can promote more sustainable and effective urban development in the context of smart cities. Traffic movement, trash management, and energy consumption data may be collected by IoT devices and safely stored on a blockchain. After that, this data may be utilized to improve resource allocation, lessen traffic, and optimize city planning.

3.2.18 IoT and Blockchain Integration

1. **Enhanced Security:** IoT devices and data can be more safety by integration of blockchain security, which is tamper-resistant and decentralized. Blockchain technology records every exchange of data or transaction inside the Internet of Things ecosystem in a transparent and unchangeable way. By lowering the likelihood of fraudulent activity, illegal access, and data breaches, this integration strengthens the system's overall resilience.

2. **Data Integrity and Immutability:** By producing an open, verifiable record of every transaction, blockchain technology guarantees the immutability of data. Ensuring the precision and dependability of data created by the IOT is imperative, particularly in vital domains such as supply chains, healthcare, and industrial operations.
3. **Decentralization:** Blockchain-enabled Internet of Things devices may function decentralized, doing away with the requirement for a central authority. As a result, there may be fewer single points of failure, more network resilience, and greater participant trust.
4. **Visible and Auditable Transactions:** Real-time monitoring and auditing of transactions and data transfers between IoT devices is made possible by blockchain's visible and traceable nature. Increased accountability and easier regulatory compliance can result from this transparency.
5. **Automated Smart Contracts:** Based on predetermined parameters, smart contracts are self-executing code recorded on the blockchain. They can automate communications and transactions among Internet of Things devices. Automation has the potential to improve efficiency, cut out middlemen, and streamline procedures.
6. **Data Monetization and Ownership**: This technology can provide safe data monetization and give owners of Internet of Things devices direct ownership over and financial gain from the data produced by their devices. This ownership change may encourage innovation and data sharing.
7. **Interoperability:** The distributed design and established protocols of blockchain can help make a variety of IoT platforms and devices interoperable with one another, allowing for smooth communication and cooperation across disparate systems.
8. **Save Expenses and Enhanced Efficiency:** IoT blockchain also has the potential to save expenses related to data reconciliation, human labor, and middlemen. This might lead to lower costs and more efficient operations.
9. **Trust and Reliability:** In situations where devices must communicate and transact with one another independently, the transparent and secure nature of blockchain can contribute to the development of trust between parties.
10. **Immutable Device Identity**: This technology may provide IoT devices with a distinct and unchangeable identity, assisting in the prevention of device spoofing, counterfeiting, and illegal access.

3.2.19 Obstacles to IoT and Blockchain Integration

IoT and blockchain integration have enormous promise, but there are a few issues that need to be resolved. When deploying these technologies, scalability, energy usage, interoperability, and cost are important factors to take into account [7].

Scalability: the Internet of Things and blockchain generate massive amounts of data, it may be challenging to ensure that the blockchain can meet the scalability requirements of IoT devices. The massive transaction volumes and latency that traditional blockchains like Bitcoin and Ethereum suffer may have an effect on real-time Internet of Things applications.

Energy Efficiency: blockchain networks and Internet of Things devices can both be resource-intensive. Energy-saving techniques should be incorporated into the integration of these technologies, particularly in situations when the devices are battery-operated or situated in remote locations.

Interoperability: Different communication protocols are used by IoT devices, which are made by different manufacturers. It might be difficult to guarantee smooth integration and compatibility between various IoT devices and blockchain networks.

Consensus techniques: traditional blockchain consensus techniques like Proof of Work (PoW) or Proof of Stake (PoS) may not be appropriate for Internet of Things applications. It is imperative to develop energy-efficient, lightweight consensus techniques that are appropriate for Internet of Things devices.

Costs: Developing and sustaining an integrated IoT-blockchain system can be expensive, requiring investments in infrastructure, development, and continuous operations. Weighing the possible advantages versus these expenses is necessary.

3.3 Conclusion

Blockchain and IoT together are like a revolutionary shift that can improve industries and transform our usage of technology. When these two are linked, it helps ensure data accuracy, upholds security, and enables automation. When people and businesses work together, they may accomplish a lot of new things and become more confident in navigating the current digital environment. With the passage of time, these technologies will become increasingly more interoperable, strengthening one another. This will enable us to progress toward a future brimming with innovative concepts and the capacity to overcome obstacles.

References

1. Rahmani, M.K.I., Shuaib, M., Alam, S., Siddiqui, S.T., Ahmad, S., Bhatia, S., Mashat, A.: Blockchain-based trust management framework for cloud computing-based internet of medical things (IoMT): a systematic review. Comput. Intell. Neurosci. **2022**, 9766844 (2022)
2. Alam, S., Shuaib, M., Khan, W.Z., Garg, S., Kaddoum, G., Hossain, M.S., Zikria, Y.: Bin blockchain-based initiatives: current state and challenges. Comput. Netw. **198**, 108395 (2021)
3. Pradhan, B., Bhattacharyya, S., Pal, K.: IoT-based applications in healthcare devices. J. Healthc. Eng. **2021**, 6632599 (2021)
4. Naresh, V.S., Pericherla, S.S., Murty, P.S.R., Sivaranjani, R.: Internet of things in healthcare: architecture, applications, challenges, and solutions. Comput. Syst. Sci. Eng. **35**, 411–421 (2020)

5. Selvaraj, S., Sundaravaradhan, S.: Challenges and opportunities in IoT healthcare systems: a systematic review. SN Appl. Sci. **2**, 139 (2020)
6. Nazir, S., Ali, Y., Ullah, N., García-Magariño, I.: Internet of things for healthcare using effects of mobile computing: a systematic literature review. Wirel. Commun. Mob. Comput. **2019**, 1–20 (2019)
7. Zheng, Z., et al.: Blockchain challenges and opportunities: a survey. Int. J. Web Grid Serv. **14**(4), 352–375 (2018). https://doi.org/10.1504/IJWGS.2018.095647

Chapter 4
Cloud-Driven Platform Services as an Integration Enabler for DLT (Blockchain), IoT, and Metadata

Kirti Jainani and Pankaj Jainani

Abstract AWS, Azure, and GCP offer extensive services, driving technology-enabled solutions across diverse industries. These platforms emphasize zero-trust security, high performance, scalability, reliability, resilience, operational excellence, and cost efficiency. They support seamless integration of Distributed Ledger Technology (DLT), Blockchain, and IoT devices, while also providing advanced tools for metadata management and data governance (Google Cloud, Cloud IoT core, Google Cloud [1]). GCP's Blockchain Node Engine and IoT Core facilitate blockchain deployment and device management. Azure's Blockchain Workbench and IoT Hub enable blockchain application development and secure device communication, with Azure Digital Twins offering spatial intelligence. AWS's Managed Blockchain and IoT Core support blockchain and device connectivity, with extensive metadata management through their cloud storage solutions. For hybrid cloud implementations, example use cases include smart supply chain management using AWS IoT Core, Azure Blockchain Service, and GCP Big Query; healthcare IoT and data privacy leveraging AWS Managed Blockchain, Azure IoT Hub, and GCP Healthcare API (Microsoft in Azure blockchain service, Microsoft Azure [2]; Google Cloud in Introduction to BigQuery ML, Google Cloud [3]; Amazon Web Services in AWS whitepapers, AWS [4]); and energy sector monitoring with AWS IoT Analytics, and Azure Digital Twins compliance with regulations like GDPR, HIPAA, and PCI DSS is crucial (Databank in HIPAA cloud compliance: Key considerations for healthcare organizations and cloud providers. Databank, 2023 [5]). AWS Config, Azure Policy, and GCP Cloud Security Command Center ensure continuous compliance and threat detection (Microsoft in Azure IoT hub. Microsoft Azure [6]; Amazon Web Services in Amazon managed blockchain, AWS [7]). Avoiding antipatterns such as over-reliance on default configurations, ignoring multi-cloud strategies, and underestimating integration complexities is essential for secure and efficient deployments.

K. Jainani (✉)
Professor (Marketing), Faculty of Management Studies, Institute of Rural Management (FMS-IRM), Jaipur, Rajasthan, India
e-mail: drkirtijainani@gmail.com

P. Jainani
Principal Architect and AVP, Genpact India Pvt. Ltd., Bangalore, Karnataka, India

P. M. Jeyanthi (ed.), *Distributed Ledger Technology in Communication*, Signals and Communication Technology, https://doi.org/10.1007/978-3-032-07682-3_4

This multifaceted approach empowers organizations to innovate and excel in the digital landscape.

Keywords AWS · Azure · GCP · Blockchain · IoT · Data analytics · Cloud architecture · PaaS

4.1 Introduction

AWS, Azure, and GCP offer extensive services that drive technology-enabled solutions across diverse industry verticals. These cloud platforms are built on foundational principles of zero-trust security, high performance, scalability, enterprise-grade reliability, resilience, operational excellence, and cost efficiency. They provide robust infrastructure and platform services, enabling seamless integration of Distributed Ledger Technology (DLT), Blockchain, and IoT devices. Furthermore, their advanced tools for metadata management, cataloguing, and lineage support comprehensive data governance and the efficient handling of diverse telemetry data sources, ensuring optimized performance and security in complex digital environments. This multifaceted approach empowers organizations to innovate and excel. This chapter will briefly outline cloud as a technology enabler for designing, implementing, deploying, and operationalization business solutions.

4.2 Platform Services on Cloud

3-Big Hyperscale's offer comprehensive solutions for integrating Distributed Ledger Technology (DLT), Internet of Things (IoT), and metadata management.

4.2.1 Leveraging Microsoft Azure Services for Blockchain and IoT Solutions

To develop a solution in Microsoft Azure using blockchain, IoT, metadata analytics, and integration, you can leverage several services and technologies. **Azure Blockchain Service** is a fully managed ledger service that provides users the ability to grow and operate blockchain networks at scale in Azure. It supports Quorum, Corda, Hyperledger Fabric, and Ethereum. **Azure IoT Hub** allows you to connect, monitor, and manage billions of IoT assets. With IoT edge computing capabilities, you can run your IoT solution at scale in the cloud. Azure IoT Central is a fully managed global IoT software as a service (SaaS) solution that makes it easy to connect, monitor, and manage your IoT assets at scale. **Microsoft Purview**, a unified data governance service, helps manage and govern on-premises, multi-cloud, and SaaS data [8]. It can provide insights into data lineage, sensitivity, and usage. **Azure Logic Apps** and **Azure Functions** offer powerful integration capabilities. Logic Apps simplifies how you design and build scalable solutions for app integration, data integration, system integration, enterprise application integration (EAI), and business-to-business (B2B) communication [9]. Azure Functions is a serverless solution that allows you to write less code, maintain less infrastructure, and save costs.

4.2.2 Leveraging Google Cloud Services for Blockchain and IoT Solutions

Google Cloud's robust network and Compute Engine facilitate the seamless hosting of blockchain nodes and the development of decentralized applications (dApps). Features such as autoscaling and rapid VM boot-up enhance these processes. Additionally, Google Kubernetes Engine (GKE) is instrumental in managing containerized applications, providing a scalable infrastructure. The Blockchain Node Engine offers fully managed blockchain node hosting, significantly simplifying Web3 development by eliminating the need for infrastructure management. This is particularly beneficial for startups aiming to bootstrap their dApp projects, especially those supporting Ethereum. Google's Blockchain Remote Procedure Call (RPC) service empowers application developers by providing a streamlined method to interact with blockchain data. This service offers a cost-effective, scalable, and reliable solution that minimizes the complexities associated with managing node infrastructure. It is fully compatible with the Ethereum JSON-RPC standard, enabling developers to integrate it into their applications with minimal code changes. Enhancing dApps with Google Cloud through high-performance, secure infrastructure, leveraging automation, AI, and programmability. Startups benefit from granular control, low-latency interconnects, and robust security measures, including Identity and Access Management (IAM) and Key Management Service (KMS), which ensure efficient and secure blockchain node communication and management. Google Cloud Blockchain Analytics simplifies access to historical blockchain data via BigQuery, enabling efficient data analysis using SQL queries. By integrating blockchain data with internal datasets, developers can gain valuable insights into user behavior and business operations. BigQuery's public datasets now include several additional blockchains, such as Avalanche, Arbitrum, Cronos, Fantom (Opera), Near, Optimism, Polkadot, Polygon Mainnet, and Tron, addressing the growing demand for comprehensive data access in the Web3 domain. Vertex AI empowers startups with scalability, real-time processing, and AI expertise to fully exploit blockchain data. This includes generating synthetic data, running simulations, and creating dynamic NFTs for customer engagement. Google Cloud's machine learning tools, like BigQuery ML and the Cloud Inference API, are vital for predictive modeling and recommendation strategies, helping develop sophisticated algorithms that uncover hidden patterns and trends in blockchain data. Google Cloud's Data Catalog is a fully managed, scalable metadata management service that helps organizations quickly discover, manage, and understand their data. It offers a unified view of all datasets, with powerful tagging and search capabilities. Google Cloud offers several integration services: Cloud Pub/Sub: Enables asynchronous messaging for application and service integration. Cloud Functions: A serverless execution environment for building and connecting cloud services with code. Apigee API Platform: Manages, secures, and analyzes APIs for more complex integration scenarios.

4.2.3 Developing Solutions with Google Cloud Technologies

To develop a solution using these technologies, follow these steps:

1. **Set up and Manage IoT Devices**: Use Cloud IoT Core to set up and manage your IoT devices.
2. **Ingest and Process Data**: Utilize Cloud Pub/Sub and Cloud Functions to ingest and process data in real-time or batch mode.
3. **Store Processed Data**: Depending on the data's nature and specific use case, store it in BigQuery or Firestore.
4. **Blockchain Integration**: Set up a private blockchain network using a solution from the GCP Marketplace, integrating it with the rest of your application using Cloud Functions or Compute Engine instances.
5. **Unified Data View**: Use Data Catalog to create a unified view of all data, including IoT and blockchain transaction data. Analyze this metadata using BigQuery or other analytics tools.
6. **API Management**: Employ Apigee to manage APIs, ensuring secure and efficient communication between different solution components.

By following these steps, you can effectively leverage Google Cloud's capabilities to develop sophisticated solutions for blockchain and IoT applications. This comprehensive approach ensures a secure, scalable, and high-performance infrastructure, empowering innovative developments in the digital landscape.

4.2.4 Leveraging AWS for Blockchain and IoT Solutions

Developing a comprehensive solution using Amazon AWS that leverages blockchain, IoT, metadata analytics, and integration technologies involves a strategic selection of services tailored to your specific needs. These choices are influenced by factors such as the nature of the problem being addressed, data scale, real-time analytics requirements, and security considerations.

AWS provides **Amazon Managed Blockchain**, which supports two widely used open-source blockchain frameworks: Hyperledger Fabric and Ethereum. Depending on your use case, such as establishing a private blockchain network or creating decentralized applications (Dapps), you can select the appropriate framework: **Hyperledger Fabric**: Ideal for creating private, permissioned blockchain networks. **Ethereum**: Suitable for developing decentralized applications and public blockchain networks [2].

For IoT applications, **AWS IoT Core** facilitates the secure and straightforward interaction of devices with cloud applications and other devices. It can support billions of devices and trillions of messages, ensuring scalability. Additionally, **AWS IoT Analytics** can be used to analyze data generated by IoT devices, providing valuable insights and enhancing decision-making processes [6]. **Metadata Analytics**—Amazon Redshift, a fully managed cloud data warehouse service, allows users to perform metadata analysis using their preferred tools. To prepare and load metadata for analytics, AWS Glue, a fully managed extract, transform, and load (ETL) service, can be employed. This combination ensures efficient data processing and analysis. Integration of various AWS services and technologies can be achieved using **AWS Lambda**, a serverless compute service that executes code in response to events and automatically manages the underlying compute resources. This facilitates seamless interaction between different components of your solution. AWS offers numerous AI and machine learning services to enhance your solutions: **Amazon SageMaker**: Facilitates building, training, and deploying machine learning models. **Amazon Comprehend**: Provides natural language processing capabilities to extract insights from text data.

To ingest data from IoT devices into AWS, **Amazon Kinesis** is an effective solution. Kinesis can collect, process, and analyze real-time streaming data, enabling immediate insights and responsive actions. **Security**—AWS ensures robust security through various services: **AWS Identity and Access Management (IAM)**: Manages secure access to AWS services. **AWS Shield**: Provides protection against Distributed Denial of Service (DDoS) attacks. **AWS Key Management Service (KMS)**: Facilitates the creation and management of cryptographic keys, ensuring data protection.

Managing and orchestrating AWS services and resources can be efficiently handled using the following: **AWS CloudFormation**: Enables infrastructure as code, allowing you to model and set up your AWS resources. **AWS Systems Manager**: Assists in operational tasks and resource management. **AWS Step Functions**: Provides serverless workflows to coordinate and automate various AWS services.

AWS offers a comprehensive suite of services to address a wide range of use cases involving blockchain, IoT, metadata analytics, and integration. By leveraging these services, you can develop robust, scalable, and secure solutions that meet your business objectives.

4.3 Reference Business Solutions on Cloud PaaS Backbone

4.3.1 Smart Supply Chain Management with Microsoft Azure

Leveraging Microsoft Azure's comprehensive suite of services, you can develop a robust and efficient supply chain management solution. This solution integrates blockchain, IoT, metadata analytics, and various integration technologies to provide transparency, traceability, and efficiency across the supply chain.

Microsoft Azure Blockchain Service can be the backbone of your supply chain solution. Blockchain's decentralized and immutable nature ensures transparency and traceability throughout the supply chain. It records every transaction from procurement to delivery, enabling real-time tracking of product journeys. This not only helps in reducing fraud and counterfeiting but also fosters trust among stakeholders.

Azure IoT Hub plays a crucial role in connecting and managing IoT devices embedded across your supply chain, from inventory sensors to delivery vehicle trackers. It facilitates real-time data collection from these devices, providing insights into inventory levels, warehouse conditions, vehicle locations, and more. This visibility helps monitor supply chain operations closely, identify potential issues early, and make data-driven decisions.

Integrating your supply chain solution with Web3 enables the use of decentralized applications (DApps) and smart contracts. Smart contracts, powered by blockchain, can automate various supply chain processes such as order placements, payments, and delivery confirmations. This automation reduces manual efforts and minimizes errors.

Azure Synapse Analytics, combined with Azure Databricks, provides powerful data analysis capabilities. These platforms can handle the vast amounts of data generated by your supply chain, offering valuable insights into trends, patterns, and anomalies. These insights can be utilized to optimize processes, predict demand, manage inventory efficiently, and formulate effective strategies.

Microsoft Fabric provides a secure, scalable, and efficient infrastructure for your supply chain solution. It is capable of managing large transaction volumes and complex operations, ensuring the smooth functioning of your supply chain. Additionally, it offers flexibility to adapt to changing business needs and scalability to grow with your business.

Microsoft Purview, a unified data governance service, helps manage and govern supply chain data. It ensures data privacy compliance, provides data lineage visibility, and enables secure data sharing with partners. Purview also assists in creating a single, reliable source of data for your supply chain, improving data accuracy and consistency.

4.3.2 Key Components of Smart Supply Chain Management

1. **Data Collection and Ingestion**: Azure IoT Hub facilitates the placement of IoT sensors throughout the supply chain on items, vehicles, and warehouses to collect real-time data. These sensors can monitor various parameters, including location, temperature, vibration, and humidity, depending on supply chain needs.
2. **Data Processing and Analytics**: After data collection, it needs to be processed and analyzed. This can be efficiently done using Azure Synapse Analytics, Databricks, and Microsoft Fabric. These platforms process large volumes of data quickly, provide insights, and enable data-driven decision-making.
3. **Blockchain for Supply Chain Transparency**: Azure Blockchain Service can be utilized to build a secure and transparent supply chain network. Each transaction or movement in the supply chain can be recorded on the blockchain, providing full traceability and preventing fraud. Smart contracts can automate various processes, such as payments and confirmations.
4. **Data Governance and Security**: Microsoft Purview ensures data governance across the entire data estate. It helps organizations discover, understand, and manage their data assets, unifying data security, governance, and compliance solutions across a heterogeneous data estate.
5. **Integration and Orchestration**: Azure Logic Apps can integrate and orchestrate data flow between different systems and services. It can connect to various data sources, perform data transformations, and trigger actions based on specific conditions.
6. **AI and Machine Learning**: Microsoft Azure offers AI and ML services to enhance supply chain operations. For instance, Azure Machine Learning can create predictive models for demand forecasting, while Azure Cognitive Services can enable features like anomaly detection in supply chain data.
7. **Presentation Layer**: Power BI can visualize data and insights derived from the system. This enables the creation of dashboards and reports that provide an at-a-glance view of the supply chain's performance.
8. **Data Storage**: Azure offers various options for data storage based on needs. Azure Data Lake Storage is suitable for big data analytics workloads, Azure SQL Database for relational data, and Azure Cosmos DB for globally distributed, multi-model database services.

4.3.3 Implementing a Secure and Efficient Healthcare Data Management Solution with Amazon AWS

The healthcare company operates a fleet of IoT diagnostic devices that generate substantial amounts of patient data. The primary objective is to ensure the security and privacy of this sensitive data while enabling efficient analysis to gain valuable healthcare insights. To meet these requirements, a hybrid blockchain solution using Amazon Managed Blockchain, Hyperledger Fabric, and Ethereum will be implemented.

Hybrid Blockchain Solution—Amazon Managed Blockchain provides a fully managed service that facilitates the creation and management of scalable blockchain networks using popular open-source frameworks such as Hyperledger Fabric and Ethereum. **Hyperledger Fabric** framework will be used for internal data that needs to be shared among the healthcare organization's departments. Hyperledger Fabric ensures secure and private data sharing within the organization. **Ethereum** framework will be utilized for external data sharing with other healthcare organizations and insurance companies. Ethereum facilitates secure and transparent interactions with external entities. This hybrid blockchain model ensures data privacy by providing each network participant with a copy of the blockchain, thus enabling them to participate in the validation of transactions.

IoT Device Integration—The IoT diagnostic devices will connect to AWS IoT Core, a managed cloud service that allows connected devices to interact easily and securely with cloud applications and other devices. **AWS IoT Core**: Supports billions of devices and trillions of messages, enabling reliable and secure processing and routing of messages to AWS endpoints and other devices.

Data Ingestion and Pre-Processing—Data collected from the IoT devices will be ingested into AWS IoT Analytics for pre-processing and transformation.

AWS IoT Analytics: A fully managed service that simplifies running and operationalizing sophisticated analytics on massive volumes of IoT data without the associated cost and complexity.

Data Storage and ETL—The cleansed data will be stored in Amazon Redshift, a fully managed, petabyte-scale data warehouse, facilitating cost-effective data analysis using standard SQL.

Amazon Redshift: Provides scalable and cost-effective data storage and analysis.

AWS Glue: A fully managed extract, transform, and load (ETL) service that prepares and loads data for analytics, catalogs the data, and makes it available for querying and analysis.

Serverless Computing with AWS Lambda—AWS Lambda, a serverless compute service, will run the company's applications and services in response to events such as changes to data in an Amazon S3 bucket or updates to a DynamoDB table.

AWS Lambda: Enables the healthcare organization to build responsive applications that quickly adapt to new information.

Machine Learning and Data Analysis—Amazon SageMaker, a fully managed service, will provide developers and data scientists with the tools to build, train, and deploy machine learning (ML) models swiftly. **Amazon SageMaker**: Used to analyze healthcare data, predict patient health trends, improve patient care, and enable data-driven decisions through advanced machine learning models.

By implementing this comprehensive solution using Amazon AWS, the healthcare company can ensure the security and privacy of sensitive patient data while enabling efficient analysis for valuable healthcare insights. The integration of IoT devices, hybrid blockchain technology, advanced analytics, and machine learning models provides a robust, scalable, and secure infrastructure that enhances the overall management and utilization of healthcare data.

4.3.4 Leveraging Google Cloud for Smart Energy Management and Optimization

Efficient energy management requires robust infrastructure, real-time data processing, and advanced analytics. Google Cloud provides a comprehensive suite of services to support this, from hosting blockchain nodes to connecting IoT devices and performing predictive analytics. This chapter outlines how to leverage Google Cloud's services to build a secure, efficient, and real-time energy management solution.

Compute Engine and Network Services offer high-performance virtual machines, providing the necessary infrastructure for hosting blockchain nodes. This ensures responsive and efficient operation. Google Cloud's Network Services enhance secure and fast communication with blockchain nodes, offering granular control over networking infrastructure, including firewall rules and IP addressing. **Blockchain RPC S**ervice simplifies interactions with blockchain data. It provides a scalable, cost-effective, and reliable solution for interacting with the blockchain, which is crucial for real-time and data-intensive applications in the energy sector. **Google Cloud Blockchain Analytics** allows easy access to historical blockchain data through BigQuery. This service provides energy companies with valuable insights into transactions and contracts on the blockchain. Combining this data with internal datasets enables deeper operational insights. **BigQuery ML** leverages machine learning to build predictive models based on blockchain data. This can be used to forecast energy demand, identify trends, and optimize energy distribution, ensuring efficient energy management. **Google Cloud IoT Core** connects IoT devices within the energy infrastructure to the cloud for real-time data collection and analysis. It securely manages and ingests data from globally dispersed devices, offering a fully managed service for collecting, processing, analyzing, and visualizing IoT data in real time. **Google Cloud's Data Catalog** is a fully managed, scalable metadata management service. It helps organizations quickly discover, manage, and understand their data within Google Cloud, offering a unified view of all datasets across the organization. **Cloud Pub/Sub** is a messaging system that decouples senders and receivers, allowing secure and reliable communication among independently written applications. It facilitates real-time updates on energy consumption, generation, and other operational metrics. **Vertex AI** provides a unified machine learning platform with tools for the entire ML workflow, from data pre-processing to model deployment. It is used to train

ML models on energy data for tasks like anomaly detection, load forecasting, and predictive maintenance.

4.3.5 *Flow and Architecture*

1. **Data Collection**: IoT devices in the energy infrastructure send real-time data to Google Cloud IoT Core.
2. **Data Publishing**: The data is published to Cloud Pub/Sub, which pushes it to BigQuery for storage and analysis.
3. **Model Building**: BigQuery ML builds machine learning models based on this data.
4. **Blockchain Interaction**: Blockchain nodes hosted on Compute Engine interact with the blockchain network, facilitated by the Blockchain RPC service.
5. **Data Analysis**: Blockchain data is analyzed using Google Cloud Blockchain Analytics, and insights are stored in BigQuery.
6. **Metadata Management**: The metadata is managed using Google Cloud's Data Catalog.
7. **Advanced Analytics**: Vertex AI is used for advanced analytics and machine learning tasks.

The proposed architecture ensures secure, efficient, and real-time monitoring and optimization for the energy sector. It leverages blockchain technology for secure data management and harnesses the power of cloud computing and machine learning for optimization and predictive analysis. By integrating these technologies, energy companies can achieve enhanced operational efficiency and data-driven decision-making.

4.4 Regulations and Compliance

It is crucial to ensure that the systems comply with various regulations and governance policies. This compliance is critical for maintaining the integrity of the system, protecting sensitive data, and adhering to the legal and ethical standards of each sector. Here are the key regulations, compliance, and governance measures to consider:

1. Data Protection and Privacy Regulations: In the healthcare sector, adherence to regulations such as HIPAA (Health Insurance Portability and Accountability Act) in the US, GDPR (General Data Protection Regulation) in the EU, and the Data Protection Act in the UK is mandatory. These regulations ensure the protection of personal health information and patient data privacy. Furthermore, for the retail sector, adherence to PCI DSS (Payment Card Industry Data Security Standard) is essential for safeguarding customers' credit card information.

2. IoT Security Measures: With the deployment of IoT devices in healthcare and energy sectors, it is crucial to comply with IoT security standards such as ISO/IEC 27,001, ISO/IEC 27,032, and NIST's Cybersecurity Framework. These standards provide guidelines on implementing robust security measures, including data encryption, secure device authentication, and secure communication protocols.
3. Blockchain Regulations: The use of blockchain technology necessitates compliance with specific regulations and standards. For instance, the EU's Blockchain Observatory and Forum provide guidelines on blockchain's legal, governance, and interoperability aspects. In the US, the SEC (Securities and Exchange Commission) provides regulations concerning blockchain and its use in financial transactions.
4. Compliance with Energy Sector Regulations: For the energy sector, adherence to regulations such as NERC CIP (North American Electric Reliability Corporation Critical Infrastructure Protection) standards is necessary. These standards provide a cybersecurity framework for identifying and protecting Critical Cyber Assets to support the reliable operation of the energy grid [7].
5. Cloud Service Provider Compliance: When using services from cloud providers like AWS, Azure, and GCP, it's important to ensure that these services comply with various industry-specific regulations. For instance, AWS, Azure, and GCP all comply with HIPAA for healthcare data and PCI DSS for financial data. Also, they adhere to privacy standards like GDPR and regional regulations like the California Consumer Privacy Act (CCPA).
6. Data Governance Policies: Implementing robust data governance policies is crucial in all sectors. These policies should define who has access to what data, how data is classified and protected, and how data quality is maintained. This also includes implementing role-based access controls (RBAC), regular security audits, and risk assessments to identify and address vulnerabilities.
7. Machine Learning and AI Ethics: If your solution incorporates machine learning and AI, it's important to consider ethical guidelines such as transparency, fairness, and accountability. This also includes complying with regulations like the EU's Ethics Guidelines for Trustworthy AI.

It's important to stay abreast of the latest regulations and standards in each sector. Regular training and audits can ensure ongoing compliance and protect your systems from security threats. Manage compliance effectively, consider using compliance management tools provided by your cloud service provider, like Azure Policy, AWS Config, and Google Cloud's Compliance Reports.

4.5 Antipatterns and Pitfalls

Cloud services such as AWS, Azure, and GCP [6] offer a plethora of services to meet various business needs. However, while utilizing these services, it's important to avoid certain antipatterns and pitfalls to ensure the efficiency and effectiveness of the solutions implemented. Here are some of the common antipatterns and pitfalls with respective mitigation strategies:

1. Challenges in Blockchain Technology: Despite its transformative potential, blockchain technology faces significant challenges related to node infrastructure, security, and data management. Operating blockchain nodes requires technical expertise to ensure resilience against data loss and to maintain service availability. Security is paramount due to the financial implications of blockchain applications, necessitating robust private key management. Furthermore, the extensive data stored on blockchains presents scalability and retrieval challenges that must be addressed for broader adoption.
2. Monolithic Architecture: Avoid building applications as a single, indivisible unit, which makes updates and scaling challenging. Instead, use a microservices architecture where each service is built around a specific business capability, independently deployable by a fully automated deployment machinery.
3. Improper Resource Allocation: Over-provisioning can lead to unnecessary costs, and under-provisioning can lead to inadequate performance. Use autoscaling features to dynamically adjust resources based on demand.
4. Neglecting Security [3]: Security should be a priority from the beginning. Ensure data encryption, use IAM roles for access control, and follow the principle of least privilege. Regularly update and patch systems to protect against vulnerabilities.
5. Lack of Disaster Recovery Plan: Always have a disaster recovery plan in place. Regularly back up data and use multi-region deployment to ensure availability in case of a region-wide outage.
6. Ignoring Costs: Cloud services can be expensive if not managed properly. Regularly monitor and optimize costs. Use cost management tools provided by cloud providers.
7. Not Considering Network Latency: Network latency can significantly impact application performance. Design your architecture keeping latency in mind. Use edge locations, Content Delivery Networks (CDNs), and local zones to reduce latency.
8. Ignoring Compliance Requirements: Ensure your application meets all the relevant compliance requirements like GDPR, HIPAA, etc. Use services like AWS Artifact, Azure Compliance Manager, and Google Cloud's Security [2] Command Center for compliance management.
9. No Monitoring and Logging: Implement comprehensive monitoring and logging from the outset. Use services like CloudWatch, Azure [1] Monitor, and Stackdriver for real-time monitoring and logging.
10. Tight Coupling with a Single Cloud Provider: Avoid vendor lock-in by designing your application to be cloud-agnostic as much as possible. This way, you maintain the flexibility to switch providers if needed.
11. Ignoring DevOps Best Practices: Incorporate DevOps practices like Infrastructure as Code (IaC), Continuous Integration and Continuous Deployment (CI/CD), and automated testing to increase efficiency and reduce errors.

In conclusion, each cloud service has its own set of potential antipatterns and pitfalls. By being aware of these and taking steps to avoid them, you can ensure that your cloud solutions are robust, efficient, secure, and cost-effective.

4.6 Conclusion

The integration of advanced cloud services from Microsoft Azure, Amazon AWS, and Google Cloud offers robust, secure, and scalable solutions across various industries. By leveraging blockchain technology, IoT, and machine learning, businesses can enhance transparency, efficiency, and decision-making processes, leading to improved operational outcomes and innovative advancements in their respective fields.

Adherence to these regulations ensures the security and integrity of your systems, fostering trust and reliability. Regular training, audits, and leveraging compliance management tools from cloud service providers can help maintain ongoing compliance, protect against security threats, and uphold ethical standards in data management and AI applications.

Awareness and proactive management of the antipatterns can lead to more efficient, secure, and cost-effective cloud solutions. Regularly revisiting best practices and leveraging available tools can significantly enhance your cloud architecture's effectiveness and resilience.

References

1. Google Cloud: Cloud IoT Core. Google Cloud (n.d.). https://cloud.google.com/iot-core
2. Microsoft: Azure Blockchain Service. Microsoft Azure (n.d.). https://azure.microsoft.com/en-us/services/blockchain-service/
3. Google Cloud: Introduction to BigQuery ML. Google Cloud (n.d.). https://cloud.google.com/bigquery-ml/docs/bigqueryml-intro
4. Amazon Web Services: AWS Whitepapers. AWS (n.d.). https://aws.amazon.com/whitepapers
5. Databank: HIPAA Cloud Compliance: Key Considerations for Healthcare Organizations and Cloud Providers. Databank (2023). https://www.databank.com/hipaa-cloud-compliance-key-considerations-for-healthcare-organizations-and-cloud-providers/
6. Microsoft: Azure IoT Hub. Microsoft Azure (n.d.). https://azure.microsoft.com/en-us/services/iot-hub/
7. Amazon Web Services: Amazon Managed Blockchain. AWS (n.d.). https://aws.amazon.com/managed-blockchain/
8. Microsoft: Microsoft Purview. Microsoft Azure (n.d.). https://azure.microsoft.com/en-us/services/purview/
9. Microsoft: Azure Logic Apps. Microsoft Azure (n.d.). https://azure.microsoft.com/en-us/services/logic-apps/
10. Microsoft: Azure IoT Central. Microsoft Azure (n.d.). https://azure.microsoft.com/en-us/services/iot-central/
11. Microsoft: Azure Functions. Microsoft Azure (n.d.). https://azure.microsoft.com/en-us/services/functions/

12. Microsoft: Introduction to Azure Service Fabric. Microsoft Docs (n.d.). https://docs.microsoft.com/en-us/azure/service-fabric/service-fabric-overview
13. Microsoft: Microsoft Security. Microsoft (n.d.). https://www.microsoft.com/en-us/security
14. Google Cloud: Compute Engine. Google Cloud (n.d.). https://cloud.google.com/compute
15. Google Cloud: Network Intelligence Center. Google Cloud (n.d.). https://cloud.google.com/network-intelligence-center
16. Google Cloud: Introducing Blockchain RPC Service for Web3 Builders. Google Cloud Blog (2022). https://cloud.google.com/blog/products/financial-services/introducing-blockchain-rpc-service-for-web3-builders
17. Google Cloud: Blockchain ETL: Public Datasets. Google Cloud Blog (2018). https://cloud.google.com/blog/products/data-analytics/blockchain-etl-public-datasets
18. Google Cloud: Machine Learning with BigQuery's New Data Types. Google Cloud Blog (2021). https://cloud.google.com/blog/products/data-analytics/machine-learning-with-bigquerys-new-data-types
19. Google Cloud: Data Catalog. Google Cloud (n.d.). https://cloud.google.com/data-catalog
20. Google Cloud: Pub/Sub. Google Cloud (n.d.). https://cloud.google.com/pubsub
21. Google Cloud: Simplifying Data Ingestion with Google Cloud's Pub/Sub BigQuery Subscriptions. Google Cloud Blog (2020). https://cloud.google.com/blog/products/data-analytics/simplifying-data-ingestion-with-google-clouds-pub-sub-bigquery-subscriptions
22. Analytics Insight: Google's Vertex AI: Features and Use Cases. Analytics Insight (2023). https://www.analyticsinsight.net/googles-vertex-ai-features-and-use-cases
23. Amazon Web Services: Amazon VPC. AWS (n.d.). https://aws.amazon.com/vpc/
24. Amazon Web Services: AWS Transit Gateway. AWS (n.d.). https://aws.amazon.com/transit-gateway/
25. Amazon Web Services: AWS PrivateLink. AWS (n.d.). https://aws.amazon.com/privatelink/
26. Amazon Web Services: AWS Direct Connect. AWS (n.d.). https://aws.amazon.com/directconnect/
27. Amazon Web Services: Introducing AWS Gateway Load Balancer. AWS News Blog (2020). https://aws.amazon.com/blogs/networking-and-content-delivery/introducing-aws-gateway-load-balancer/
28. Amazon Web Services: AWS Network Firewall: New Managed Firewall Service in VPC. AWS News Blog (2020). https://aws.amazon.com/blogs/aws/aws-network-firewall-new-managed-firewall-service-in-vpc/
29. Amazon Web Services: Amazon Route 53. AWS (n.d.). https://aws.amazon.com/route53/
30. Amazon Web Services: AWS Cloud Adoption Framework. AWS Solutions (n.d.). https://aws.amazon.com/solutions/aws-cloud-adoption-framework/
31. Amazon Web Services: AWS Well-Architected Framework. AWS Documentation (2020). https://docs.aws.amazon.com/wellarchitected/latest/framework/wellarchitected-framework.pdf
32. Amazon Web Services: Amazon S3. AWS (n.d.). https://aws.amazon.com/s3/
33. Microsoft: Introduction to Microsoft Cloud for Healthcare reference architectures. Microsoft Docs (n.d.). https://docs.microsoft.com/en-us/learn/modules/intro-microsoft-cloud-healthcare-reference-architectures/
34. Ethereum Foundation: Web3.js v1.5.2 Documentation. Read the Docs (n.d.). https://web3js.readthedocs.io/en/v1.5.2/
35. A3Logics: AI in Supply Chain. A3Logics (2023). https://a3logics.com/blog/ai-in-supply-chain

Chapter 5
Absorption of Metadata, IoT, and Blockchain for Security Smart Contacts Communication Between the Hospital Industry and Patients

Bhupinder Singh, Christian Kaunert, and Saurabh Chandra

Abstract Smart contracts are a set of commands that can be used to carry out any transaction on the metadata, IoT, and distributed ledger technology. These smart contracts arrange solidity programming language to program any kind of activity on it. Metadata is critical in guaranteeing the accuracy and accessibility of health data. This section goes into how metadata may be strategically used to improve security measures in hospital-patient communication by studying successful implementations. Real-world case studies demonstrate metadata's usefulness in data management, leading to a strong and secure communication framework. The Internet of Things (IoT) has transformed healthcare by providing real-time health monitoring. This section delves into linked healthcare devices and their uses in hospitals and households. It examines the security protections built into IoT devices, addressing privacy concerns as well as regulatory problems in the context of health care. The chapter investigates how blockchain, in conjunction with Smart Contracts, might establish an immutable and safe environment for the sharing of health information. Real-world examples provide insights, underlining the potential of this integration to protect patient data and expedite communication procedures. This chapter comprehensively explores assimilation of metadata, IoT, distributed ledger technology, and blockchain represents a cutting-edge approach to enhancing security in hospital-patient communication in smart contracts.

Keywords Metadata · Smart contracts · Confidential communications · Privacy · Legal-ethical aspects

B. Singh (✉)
Sharda University, Greater Noida, India
e-mail: bhupindersinghlaw19@gmail.com

C. Kaunert
Dublin City University, Dublin, Ireland

University of South Wales, Cardiff, UK

S. Chandra
School of Law, Bennett University, Greater Noida, India

P. M. Jeyanthi (ed.), *Distributed Ledger Technology in Communication*, Signals and Communication Technology, https://doi.org/10.1007/978-3-032-07682-3_5

5.1 Introduction

The health sector has leveraged Internet of Things (IoT) technology to develop numerous services, enabling prompt patient treatment by streamlining drug supply and enhancing patient monitoring [1]. With optimizing medication delivery and improving patient monitoring, the health industry has made extensive use of Internet of Things (IoT) technology to provide a range of services that facilitate timely patient care [2, 3]. This invention also applies to specialized care for senior citizens, speeding up data collection for physicians to aid in quicker diagnosis and treatment [4]. However, there are security issues with IoT technology, especially when it comes to sensitive data. The Internet of Things and Blockchain technologies have recently addressed a number of security issues [5].

Blockchain increases efficiency and improves health services by making it more difficult to tamper with medical data or hack IoT systems [6]. Blockchain is being hailed as a promising approach to protect privacy and security in edge computing networks, as well as a way to advance edge computing technology in the future [7]. The combination of blockchain technology and edge computing ushers in a new age for edge-IoT networks, bringing revolutionary shifts that make way for new industrial and consumer services and applications [8]. But blockchain's high processing power poses a challenge for integrating it with IoT, which is constrained by space and resource limitations [9]. Because of its high degree of security and transparency, it may be used for a variety of purposes in the healthcare sector, which can save a lot of money and create new methods for patients to get treatment [10]. This future-proof technology may further propel an age of growth and innovation by combining data and innovation. Its transformational potential is already being laid out by forward-thinking enterprises and the sub sections of introduction sections shows in Fig. 5.1 [11].

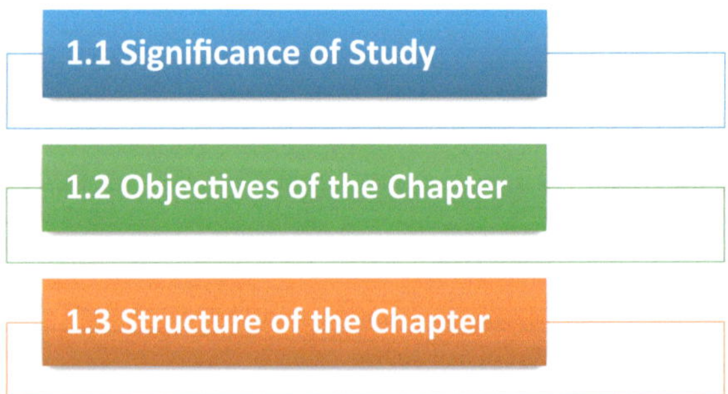

Fig. 5.1 The landscapes of introduction split sections (*Source*—Original)

5.1.1 Significance of the Study

Blockchain technology has the power to completely transform the healthcare industry's IT infrastructure, transforming it from a small-scale, isolated, and centralized system to a decentralized, international one [12]. This change has the potential to greatly raise care quality while getting rid of expensive administrative inefficiencies [13]. It may make financial transactions between patients and healthcare institutions easier, especially when it comes to micropayments [14]. The remuneration of healthcare practitioners may change in evolving value-based healthcare models from being focused on quantity to well-being [15]. There is a lot of room for innovation when IoT devices designed specifically for the healthcare industry are widely adopted [16]. The vast amount of data produced by these networked devices has the potential to completely transform healthcare [17]. Data is collected or processed at one level and then fed back into the next, forming a coherent system [18]. Value integration promotes insights and opens up dynamic business prospects throughout this process [19].

5.1.2 Objectives of the Chapter

The goal of this study is to develop a blockchain-based system that will improve patient data administration and security [20]. This solution creates an accessible, safe, decentralized healthcare environment by utilizing smart contracts and blockchain technology [21]. The people will now be able to freely and securely exchange their medical records with physicians, hospitals, research institutes, and other organizations while maintaining complete control over the privacy of their personal health information [22, 23]. This chapter has the following objectives, as shown in Fig. 5.2:

- Develop metadata, blockchain, and IoT-based system for management and security of data.
- Establish accessible and decentralized health records and communication of patient's data in a confidential manner.
- Examine the role of metadata in health data communication.
- Explore the integration of IoT for real-time health monitoring.
- Investigate the application of Blockchain and Smart Contracts for secure health information exchange.

5.1.3 Structure of the Chapter

This chapter deeply dives into the Absorption of Metadata, IoT, and blockchain for Security Smart Contacts Communication between Hospital Industry and Patients:

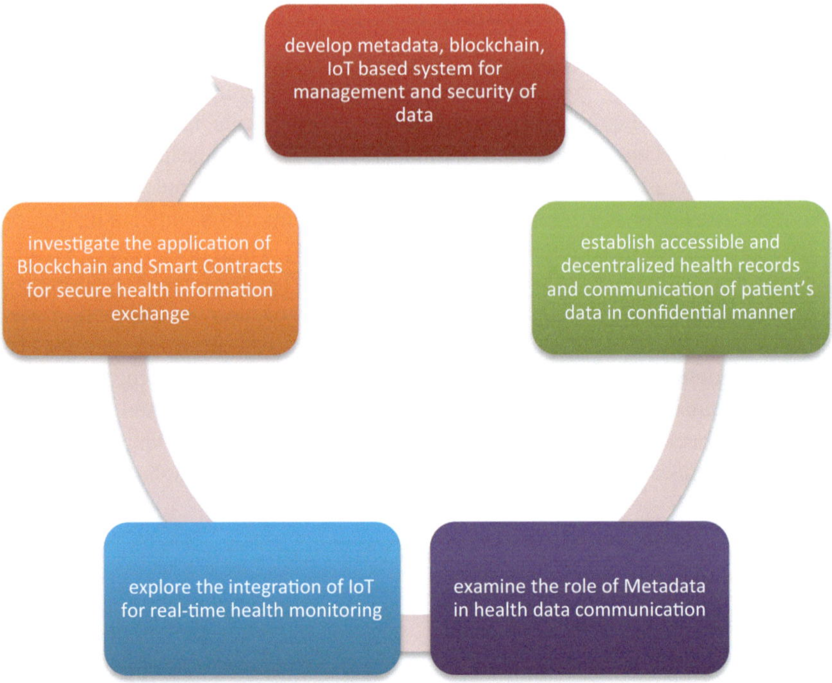

Fig. 5.2 Express the objectives of the chapter (*Source*—Original)

Lensing State-of-the-art Distributed Ledger Technology (DLT) in Machine-to-Machine for Health Information Exchange Projecting Futuristic Dimensions. Section 5.2 elaborates the Metadata in Healthcare Communication: Relevance and Importance. Section 5.3 expresses the IoT in Real-Time Health Monitoring: Connected Healthcare Devices. Section 5.4 discusses the Blockchain and Smart Contracts in Health Information Exchange. Section 5.5 lays down the Regulatory Compliance and Ethical Implications. Section 5.6 Conclude the Chapter with Future Scope as shows in Fig. 5.3.

5.2 Metadata in Healthcare Communication: Relevance and Importance

Every patient encounter and diagnosis generates enormous volumes of data in the complex and dynamic healthcare industry [24]. It can be difficult to handle this data, which emphasizes the need for metadata management [25]. With giving the information more context, meaning, and structure, this method helps firms manage their data [26]. Healthcare companies may improve patient care and research outcomes by

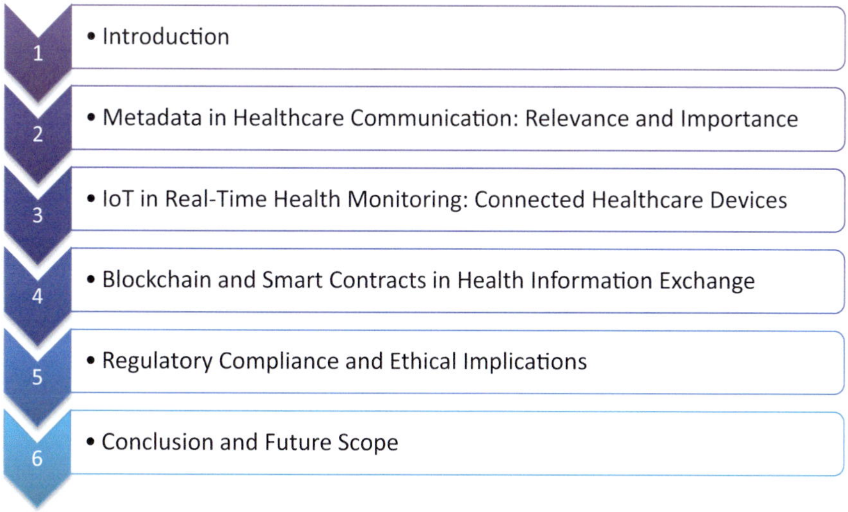

Fig. 5.3 The flow of this chapter (*Source*—Original)

optimizing their data consumption via the deployment of effective metadata management solutions [27]. Metadata, often referred to as "data about other data" serves various purposes, whether it's describing the content of a webpage, providing technical details of an image, or specifying usage rights for an asset, as shown in Fig. 5.4 [28, 29].

Fig. 5.4 The metadata in healthcare communication: relevance and importance (*Source*—Original)

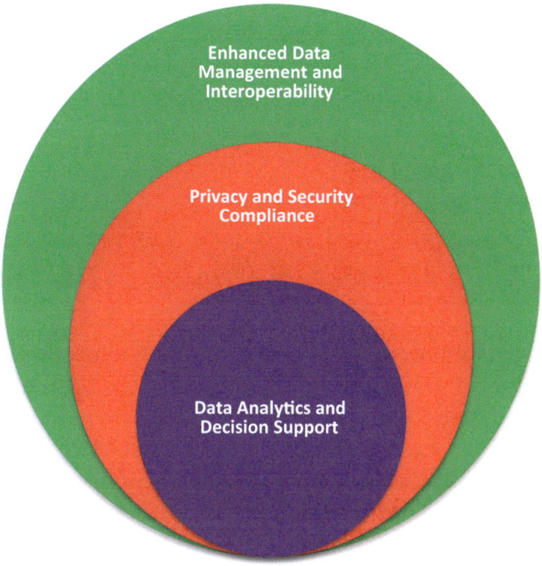

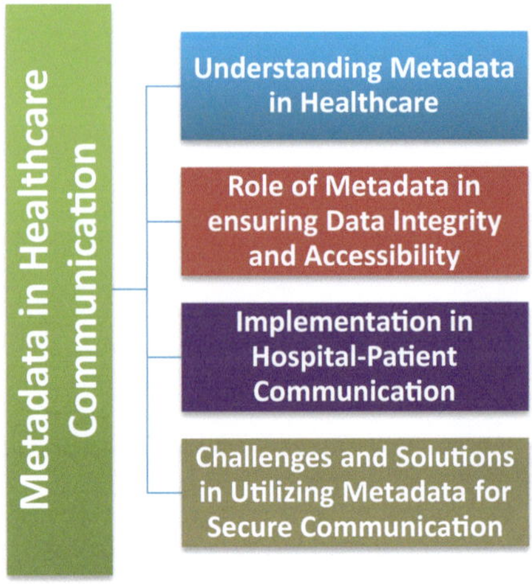

Fig. 5.5 Express the main genesis of metadata in healthcare communication (*Source*—Original)

Metadata facilitates data management by providing more information, which makes it possible to discover and use assets more efficiently [30]. Metadata is primarily intended for machine reading; it is not visible to the average web user. For example, information attached to a webpage helps search engines understand and appropriately classify pages [31]. Analogously, information encoded in a picture guarantees accurate classification and makes it simple to locate in a digital asset management (DAM) system. Also, metadata is important for identifying use rights, promoting material reuse, and sharing information [32]. With metadata, clinical decision-making may be supported by more accurate and trustworthy data, which can enhance patient care [33]. For example, metadata helps identify patients who are more likely to experience difficulties or readmission, which helps healthcare practitioners create more efficient treatment plans [34]. This metadata guarantees that patients receive appropriate and timely treatments and promotes improved coordination and communication between healthcare professionals, as shown in Fig. 5.5 [35].

5.2.1 Understanding Metadata in Healthcare

In the healthcare industry, administrative, structural, descriptive, and technological metadata are the four main categories of metadata [36]. The ownership, access permissions, and other administrative details of healthcare data are defined by administrative metadata. Database schemas and table structures are examples of the structural metadata that describe the organization of healthcare data. Information about healthcare data that helps to describe its content, context, and structure is referred to as metadata in the field of medicine [37]. Data definitions, field names, data sources, information on data quality, and other technical and administrative elements are included in this. In order to maintain the accuracy, dependability, and accessibility of healthcare data for all stakeholders, metadata is essential to its management and analysis [38, 39].

5.2.2 Role of Metadata in Ensuring Data Integrity and Accessibility

These tags, collectively known as metadata, are often overlooked by casual users, but they are essential for improving the visibility of a website on search engines and social media. Furthermore, there are several types of information in our surroundings [40]. For example, almost all contemporary digital cameras gather technical metadata for every shot, which includes information like the location, the time and date of the shot, and the size of the image. Also, structural metadata, which provides information on a document's structure or an asset's usage is a crucial component of content management systems and digital asset management systems. Whatever its form, metadata's core objective is always the same: to help companies find and use their resources more efficiently in the middle of ever-growing amounts of data [41].

5.2.3 Implementation in Hospital-Patient Communication

Metadata plays a critical role in improving the quality and usability of data in the healthcare industry. It gives the data context and meaning, making comprehension and analysis simpler. Furthermore, metadata guarantees data integrity, quality, and completeness, which promotes improved integration and sharing that is essential for improving patient care and research [42]. A number of advantages come from efficient metadata management in the healthcare industry, such as better decision-making, more accurate analysis, higher-quality data, and simpler access for stakeholders. Additionally, it reduces time and effort by streamlining data management procedures. Information about the nature and importance of healthcare data is provided by descriptive metadata, which includes field names and data descriptions. Technical information, such as file type, size, and encoding, provides light on the technical features of healthcare data [43].

5.2.4 Challenges and Solutions in Utilizing Metadata for Secure Communication

Healthcare metadata management faces a number of difficulties despite its advantages. Effective integration and management are hampered by data silos, which are compartmentalized data spread across several departments and systems. Another issue is that legacy systems were not intended for metadata management. In complicated healthcare contexts, establishing clear ownership and accountability for data is essential, but it may be challenging [44]. Finally, in order to successfully use new metadata management techniques, healthcare personnel might need assistance and training. There are several approaches that businesses may use to address metadata management issues in the healthcare industry. Setting up explicit metadata governance procedures that define who is responsible for what healthcare data is is an efficient strategy. Another helpful step is to implement metadata repositories, which provide a centralized platform for organizing and storing metadata. Moreover, improving metadata consistency and quality across many systems and departments may be achieved through the creation of metadata models that standardize content and structure [45].

5.3 IoT in Real-Time Health Monitoring: Connected Healthcare Devices

IoT devices give several chances for patients to monitor themselves and for healthcare providers to track their patients, providing advantages and difficulties for both sides. The main uses of Internet of Things devices in healthcare are remote patient monitoring. These devices eliminate the need for patients to travel or collect data on their own by automatically gathering vital signs like blood pressure, temperature, and heart rate from patients who are not physically present at healthcare institutions. IoT devices provide their acquired data to software programs so that patients and/or healthcare providers may access it. This data may be analyzed by algorithms to suggest treatments or produce alarms [46]. Assuring the security and privacy of the extremely private information gathered by remote patient monitoring devices is a substantial barrier, nevertheless. Particularly helpful is glucose monitoring for the 30 million Americans who suffer from diabetes. IoT devices provide continuous, automated glucose monitoring, doing away with the need for human record-keeping and sending out notifications when blood sugar levels become dangerous. One of the challenges is creating devices that do not need to be recharged often and are tiny enough to monitor continuously without causing any disruptions, as shown in Fig. 5.6 [47].

IoT devices make heart rate monitoring easier and provide continuous monitoring without limiting patient movement. Even with current technology, obtaining extremely exact findings is still difficult, yet 90% accuracy is achievable with most instruments. IoT devices that remind people to clean their hands upon entering improve the monitoring of hand hygiene in healthcare institutions [48]. But these gadgets aren't able to cleanse hands; they can only serve as reminders. "Mood-aware" Internet of Things (IoT) devices make depression and mood monitoring easier by gathering and analyzing data, such blood pressure and heart rate, to infer

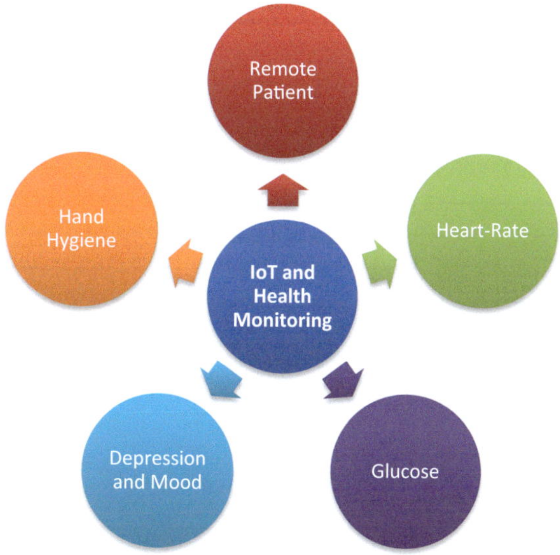

Fig. 5.6 The key components of IoT and health monitoring (*Source*—Original)

a patient's mental state [49]. The difficulties are that depression symptoms cannot be completely accurately predicted. IoT devices that continually gather data about symptoms simplify the monitoring of Parkinson's disease and enable patients to continue with their regular activities at home rather than needing prolonged hospital visits for observation as shows in Fig. 5.7 [50].

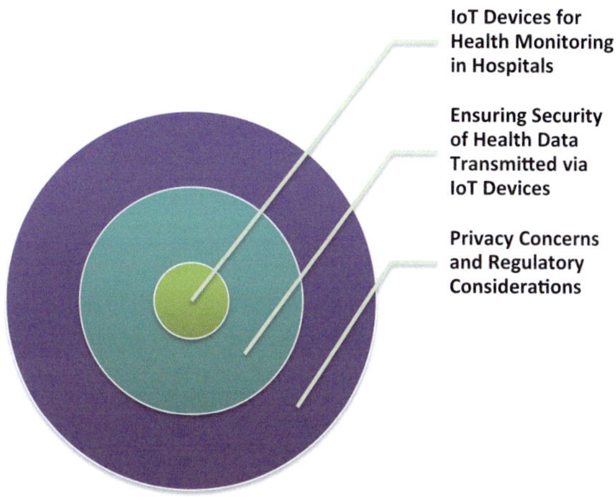

Fig. 5.7 Illuminates the sub-sections of IoT in real-time health monitoring (*Source*—Original)

5.3.1 *Overview of IoT Devices for Health Monitoring in Hospitals*

Ensuring high-quality patient care is the goal of this integrated healthcare system [51]. Vital parameters are monitored via sensors, and a WiFi module sends the data these sensors collect to the cloud [52]. As a result, a wireless healthcare monitoring system that can give online, real-time information on a patient's status has been developed. Over the past ten years, healthcare monitoring systems have undergone tremendous evolution and have become more technologically sophisticated [53]. Regretfully, many people experience a variety of health problems and early mortality as a result of untreated diseases. An Internet of Things-based real-time health monitoring system has been suggested as a solution to this problem [54]. This system includes a GSM and a mobile application for ongoing wireless patient monitoring. The main goal is to implement an IoT-based dependable patient management system that will allow medical staff to keep an eye on patients at home or in a hospital [55].

5.3.2 *Data Security and Privacy-Ensuring Security of Health Data Transmitted via IoT Devices*

Internet of Things (IoT) healthcare systems face a number of security issues, including data persistence, device authentication, confidentiality, integrity, and privacy. There are several concerns associated with data theft, privacy violations, and confidentiality for environmental nodes such as sensors, actuators, and gateways [56]. Attackers take advantage of holes in home networks to steal data that may be sold for a profit. Data profiling is another common issue with healthcare IoT systems, as attackers try to use stolen data to damage specific people. Attackers use a variety of strategies to interfere with Internet of Things systems, which emphasizes the need for strong security measures to prevent unwanted activity [57]. IoT security is still a major worry despite its fast expansion, especially in industries like healthcare, where IoT applications are expanding. Security is crucial since healthcare systems handle sensitive patient data and other personal data by nature [58]. Because IoT depends on internet access, data collected by sensors may be compromised. Strong protected communication protocols are required for large-scale IoT implementation security. To avoid data theft and profiling, it is also essential to protect the databases that hold this information [59].

5.3.3 Privacy Concerns and Regulatory Considerations in IoT Implementation

This smart contract is built on the blockchain and can handle a number of circumstances, such as data access and permission administration. This arrangement involves several parties, each of whom has been given a distinct duty, and it helps to improve communication between medical professionals and patients [60]. Smart contracts may help monitor all the activities related to a unique ID from creation to submission and include restrictions for data permission [61]. A number of situations have been described and explained, as well as the operations and processes included in the smart contracts [62]. Through the smart contract, the operation may be immediately monitored and approved, removing the need for involvement from a centralized authority and drastically lowering administration expenses [63]. All medical record data is kept in a local database storage to maximize efficiency [64]. Most of the project's goals have been effectively met by the system, especially the safe blockchain-based exchange of medical records, which has satisfied user demands [65]. As a result, blockchain has been successfully used to carry out medical report verification, data sharing, and security [66]. Also, with lowering reliance on physical copies of medical information and minimizing dangers related to possible human transmission chains, the system solves issues related to disease transmission in hospitals, such as the continuing COVID-19 epidemic [67].

5.4 Blockchain and Smart Contracts in Health Information Exchange

Through the integration of powerful computers, sensors, and actuators, the Internet of Things (IoT) links people, things, and devices to enable data gathering [68]. All of these components broadcast data to a central host, which is usually a cloud server [69]. Data gathered from IoT sources may be analyzed to produce insights and methods that can improve current services and have an influence on corporate operations [70]. However, there are major obstacles to the IoT ecosystem's security and privacy, which prevent mainstream adoption. IoT systems are regularly the target of ransomware, DDoS attacks, and other malicious intrusions [71]. DDoS attacks are especially dangerous because they overwhelm a target, like a central database, with a barrage of simultaneous data requests from multiple compromised network systems, limiting access for impacted users [72, 73]. Current centralized systems may run into difficulties recognizing, authenticating, and integrating new nodes into the network as the number of devices accessing IoT networks rises [74].

Blockchain, often referred to as Distributed Ledger Technology (DLT), emerges as a game-changing invention that can solve a number of problems pertaining to IoT security, privacy, and scalability in order to meet these IoT concerns [75]. The immutable ledger of blockchain technology guarantees records that cannot be tampered with, removing the need for third parties to verify them [76]. The deployment of IoT is shown by a variety of applications, such as smart communities, architecture, grids, mobility, housing, and medical facilities [77]. For those in need, eHealth provides quick access to clinical treatments and other advantages [78]. Blockchain technology has the potential to significantly improve overall efficiency to support patient care and enable the gradual conversion of current medical systems into dispersed eHealth platforms. It can also effectively address important security and confidentiality concerns [79, 80].

5.4.1 Benefits of Blockchain in Ensuring Transparency and Security in Health Data Transactions

A new industry known as Blockchain for IoT (BCIoT) is created when blockchain technology is integrated into the Internet of Things [81]. This massive volume of data produced by IoT devices benefits from the consistent access worldwide under the BCIoT concept [82]. There are numerous parties that can access data obtained from activities via smart contracts [83]. The data validity inside the blockchain-based network is guaranteed by telecommunications carriers and other third parties [84]. Nonetheless, security and privacy continue to be the top priorities in any blockchain network, which might undermine user confidence and flexibility [85].

5.4.2 Smart Contracts for M2M Communication

A collection of instructions used to carry out transactions on the blockchain is referred to as a smart contract [86]. This section of code is active once users start transactions [87, 88]. It works directly with the blockchain; thus, it cannot be altered or tampered with [89]. Smart contracts describe different blockchain activities using the solidity programming language. Developers can compile and perform these actions on the blockchain after they have been written [90]. Solidity is a programming language that is often implemented using JavaScript to write smart contract code [91]. A decentralized data storage platform with a peer-to-peer network design is called Interplanetary File System (IPFS) [92]. With giving data files cryptographic identities, IPFS provides safe data storage by limiting attempts to change stored data to changing the identifier [93, 94]. A cryptographically generated hash value, which both uniquely identifies and protects each data file saved on IPFS from modification, is assigned to each file [95]. Using peer-to-peer (P2P) connections, the IPFS protocol uses IPFS objects, which are collections of data and connections [96]. Whereas the linkage is made up of disorderly binary values, the data component is an array of disorganized binary values [97].

5.4.3 Applications of Smart Contracts in Machine-to-Machine Interactions

Machine-to-machine (M2M) systems face a number of security issues, such as device manipulation and wireless infiltration, in addition to illegal access [98]. It is important to consider many factors, including physical security, privacy issues, fraud prevention, and the protection of mission-critical applications [99]. Increasing the resistance of machines and devices to tampering, building backend servers with security features, encrypting communications, and embedding security measures directly into the machinery are common M2M security approaches [100, 101]. Also, the controlling device identities, guaranteeing data confidentiality, keeping device availability, and isolating M2M devices onto specialized networks are useful tactics for reducing M2M security threats [102]. The traditional approach to maintaining medical records is ineffective and necessitates a large amount of storage space in order to hold all of the test data for each patient. Information transfer was hampered by the unstructured data storage of earlier systems [103].

The healthcare sector generates enormous amounts of data, therefore improving data management techniques while preserving security and privacy is imperative [104]. Data secrecy requires more modifications, which creates a number of issues that blockchain successfully resolves [105]. With a simple click, patients may use this system to give or withdraw certain record authorizations to authorities [106, 107]. The implementation of this automation has been considerably expedited by utilizing Ethereum and smart contracts [108]. The suggested wallet acts as a bridge for patients who are reluctant to transfer their data to electronic health records (EHRs) and as a safe and convenient means to use blockchain technology [109]. It also streamlines the administration of secret keys [110]. The cryptographic encryption techniques used by the system provide strong security and dependability, as well as defense against unwanted access, as shown in Fig. 5.8 [111].

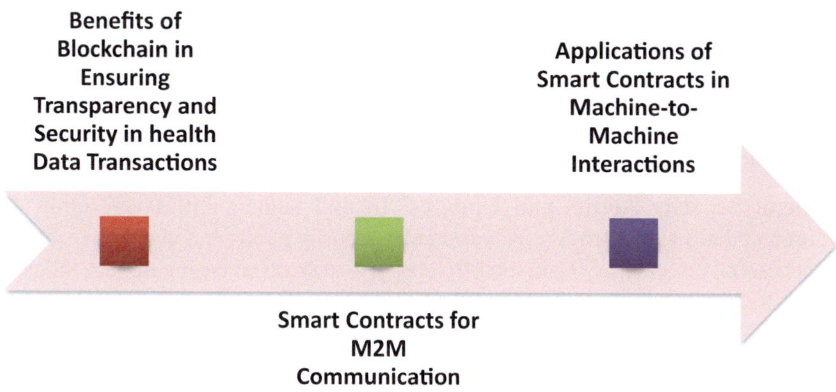

Fig. 5.8 Specifies the blockchain and smart contracts in health information exchange (*Source*—Original)

5.5 Regulatory Compliance and Ethical Implications

The sensitive data availability, confidentiality, and integrity must be given top priority in a strong cloud-based monitoring system for medical data inside the Internet of Things (IoT) ecosystem [112]. This means preserving data integrity throughout processing and storing in the cloud, protecting patient privacy, and preventing unwanted access [113, 114]. This is a summary of the goals and tactics for putting such a system into place as follows:

Data Encryption: Use strong encryption techniques to protect medical data both in transit and in storage [115]. Data encrypted is guaranteed to remain unreadable and useless even in the event of illegal access or interceptions [116].

Access Control and Authentication: Put strict controls in place to ensure that only systems or people with permission may access medical records [117]. This entails limiting data access in accordance with user responsibilities and obligations by utilizing strong authentication approaches like role-based access control (RBAC) and two-factor authentication [118].

Safe Routes: Create safe routes of communication between cloud computing infrastructure, IoT devices, and medical applications [119]. This involves encrypting data in transit to avoid tampering or eavesdropping by using secure protocols like Secure Sockets Layer (SSL) or Transport Layer Security (TLS) [120].

Validation and Data Integrity: Implement systems to guarantee the accuracy and completeness of medical records [121]. Message authentication codes (MACs) and digital signatures, which ensure that data is not changed in transit or storage, can be used to do this [122].

Security Monitoring and Logging: Put monitoring systems in place to find any unauthorized access attempts or unusual activity [123]. To find any anomalies or potential security breaches, log and examine network traffic, system events, and user activity [124].

Regulatory Compliance: Make sure that relevant healthcare laws and regulations, such as the General Data Protection Regulation (GDPR) and the Health Insurance Portability and Accountability Act (HIPAA), are followed. Respect the necessary privacy and security requirements specified by this legislation [125, 126].

Frequent Security Audits and Updates: To find vulnerabilities and apply the required updates and patches, conduct regular security audits and evaluations [127]. To successfully maintain system security, stay up to date on emerging threats, industry standards, and the newest security techniques [128].

Secure Storage: To protect medical records on the cloud, make use of secure storage options [129]. This might include using redundancy, access restrictions, and encryption when data is at rest in order to guarantee data availability and resilience against cyberattacks and hardware malfunctions, as shown in Fig. 5.9 [130].

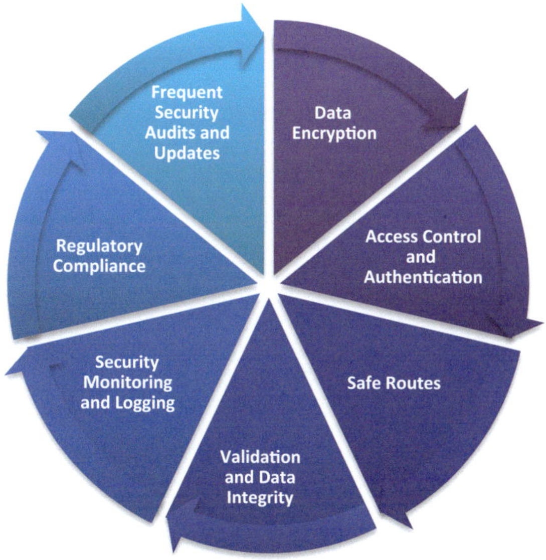

Fig. 5.9 Highlights the important points of regulatory compliance and ethical implications (*Source*—Original)

5.6 Conclusion and Future Scope

By using this technology, interoperability might be improved while maintaining data security and secrecy. It is, by its very nature, honest and follows the letter of the law. Health outcomes stand to gain from increased interoperability since it will make it easier for different players in the healthcare value chain to exchange network access without sacrificing data security. This technology makes it possible for information to be exchanged in a decentralized, transparent, safe, and reliable manner. As a result, it is expected that records be updated on a regular basis and that there be no discrepancies between the two databases, making the process of validation and reconciliation easier. This evaluation highlights the ways in which the technology's application delivers healthcare solutions and tackles issues related to scalability. It has the ability to make financial transactions between patients and healthcare institutions easier, especially when it comes to micropayments. Payment for services provided by providers may shift under future value-based healthcare models such that quality is prioritized above quantity. With the use of this technology, healthcare organizations may securely update patient data in real-time across several institutions and locations while maintaining the privacy of medical information.

References

1. Zhang, X., Sheng, Y., Liu, Z.: Using expertise as an intermediary: Unleashing the power of blockchain technology to drive future sustainable management using hidden champions. Heliyon **10**(1) (2024)
2. Khan, T., Civas, M., Cetinkaya, O., Abbasi, N.A., Akan, O.B.: Nanosensor networks for smart health care. In: Nanosensors for Smart Cities, pp. 387–403. Elsevier (2020)
3. Agoulmine, N., Kim, K., Kim, S., Rim, T., Lee, J.S., Meyyappan, M.: Enabling communication and cooperation in bio-nanosensor networks: toward innovative healthcare solutions. IEEE Wirel. Commun. **19**(5), 42–51 (2012)
4. Dorj, U.O., Lee, M., Choi, J.Y., Lee, Y.K., Jeong, G.: The intelligent healthcare data management system using nanosensors. J. Sens. **2017** (2017)
5. Rani, P.: Nanosensors and their potential role in internet of medical things. In: Nanosensors for Futuristic Smart and Intelligent Healthcare Systems, pp. 293–317. CRC Press (2022)
6. Rubeis, G.: Ethical implications of blockchain technology in biomedical research. Ethik in der Medizin, pp. 1–14 (2024)
7. Ahirwar, R., Khan, N.: Smart Wireless Nanosensor Systems for Human Healthcare, pp. 265–292. CRC Press (2022)
8. Han, B., Tomer, V., Nguyen, T. A., Farmani, A., Singh, P.K. (eds.): Nanosensors for Smart Cities. Elsevier (2020)
9. Sarkar, S., Misra, S.: From micro to nano: The evolution of wireless sensor-based health care. IEEE Pulse **7**(1), 21–25 (2016)
10. Rizwan, A., Zoha, A., Zhang, R., Ahmad, W., Arshad, K., Ali, N.A., et al.: A review on the role of nano-communication in future healthcare systems: A big data analytics perspective. IEEE Access **6**, 41903–41920 (2018)
11. Datta, S.P.A.: Future healthcare: bioinformatics, nano-sensors, and emerging innovations. Nanosens. Theory Appl. Ind. Healthcare Defense **247** (2016)
12. Pramanik, P.K.D., Solanki, A., Debnath, A., Nayyar, A., El-Sappagh, S., Kwak, K.S.: Advancing modern healthcare with nanotechnology, nanobiosensors, and internet of nano things: taxonomies, applications, architecture, and challenges. IEEE Access **8**, 65230–65266 (2020)
13. Archana, K., Gloth, A.: Blockchain and entrepreneurship. In: Applying Business Intelligence and Innovation to Entrepreneurship, pp. 35–51. IGI Global (2024)
14. Nazari, A.: Nanosensors for smart cities: an introduction. In: Nanosensors for Smart Cities, pp. 3–8. Elsevier (2020)
15. Sahi, K.S.S., Kaushik, S.: Smart nanosensors for healthcare monitoring and disease detection using AIoT framework. In: Nanosensors for Futuristic Smart and Intelligent Healthcare Systems, pp. 387–400. CRC Press (2022)
16. Oukhatar, A., Bakhouya, M., El Ouadghiri, D.: Electromagnetic-based wireless nano-sensors network: architectures and applications. J. Commun. **16**(1), 8 (2021)
17. Ramasamy, M., Kumar, P.S., Varadan, V.K.: Wearable nanosensor systems and their applications in healthcare. In: Nanosensors, Biosensors, Info-Tech Sensors and 3D Systems 2017, vol. 10167, p. 1016703. SPIE (2017)
18. Gao, Q.: Artificial Intelligence Governance and the Blockchain Revolution. Springer Nature (2024)
19. Paliwal, G., Bunglowala, A.: Nanosensor and actuator technologies for wearable mobile patient monitoring systems: a review. In: Workshop on Computational Models of Natural Argument, pp. 83–95. Springer Singapore, Singapore (2018)
20. He, L., & Eastburn, M. (2022, August). Smart nanosensor networks for body injury detection. In: 2022 IEEE International Conference on Smart Internet of Things (SmartIoT), pp. 15–19. IEEE (2022)
21. Ali, N.A., Abu-Elkheir, M.: Internet of nano-things healthcare applications: requirements, opportunities, and challenges. In: 2015 IEEE 11th International Conference on Wireless and Mobile Computing, Networking and Communications (WiMob), pp. 9–14. IEEE (2015)

22. Pirmagomedov, R., Hudoev, I., Shangina, D.: Simulation of medical sensor nanonetwork applications traffic. In: Distributed Computer and Communication Networks: 19th International Conference, DCCN 2016, Moscow, Russia, November 21–25, 2016, Revised Selected Papers 19, pp. 430–441. Springer International Publishing (2016)
23. Mujawar, M.A., Gohel, H., Bhardwaj, S.K., Srinivasan, S., Hickman, N., Kaushik, A.: Nano-enabled biosensing systems for intelligent healthcare: towards COVID-19 management. Mater. Today Chem. **17**, 100306 (2020)
24. Omanović-Mikličanin, E., Maksimović, M., Vujović, V.: The future of healthcare: nanomedicine and internet of nano things. Folia Medica Facultatis Medicinae Universitatis Saraeviensis **50**(1), 23–28 (2015)
25. Udoh, E.E., Hermel, M., Bharmal, M.I., Nayak, A., Patel, S., Butlin, M., Bhavnani, S.P.: Nanosensor technologies and the digital transformation of healthcare. Pers. Med. **20**(3), 251–269 (2023)
26. Shafik, W.: Data-driven future trends and innovation in telemedicine. In: Improving Security, Privacy, and Connectivity Among Telemedicine Platforms, pp. 93–118. IGI Global (2024)
27. Chelliah, R., Wei, S., Daliri, E.B.M., Rubab, M., Elahi, F., Yeon, S.J., et al.: Development of nanosensors based intelligent packaging systems: food quality and medicine. Nanomaterials **11**(6), 1515 (2021)
28. Topel, S.D., Al-Turjman, F.: Nanosensors for the internet of nano-things (IoNT): an overview In: Internet of Nano-Things and Wireless Body Area Networks (WBAN), pp. 21–44 (2019)
29. Dinesh kumar, K.: Human health monitoring mobile phone application by using the wireless nanosensor based embedded system. In: 2013 International Conference on Information Communication and Embedded Systems (ICICES), pp. 889–892. IEEE (2013)
30. Materon, E.M., Gómez, F.R., Joshi, N., Dalmaschio, C.J., Carrilho, E., Oliveira, Jr, O.N.: Smart materials for electrochemical flexible nanosensors: advances and applications. In: Nanosensors for Smart Manufacturing, pp. 347–371 (2021)
31. Kumar, S., Kumari, P.: Flexible nano smart sensors. Nanosens. Smart Manuf. **199** (2021)
32. Gaobotse, G., Mbunge, E., Batani, J., Muchemwa, B.: Non-invasive smart implants in healthcare: redefining healthcare services delivery through sensors and emerging digital health technologies. Sens. Int. **3**, 100156 (2022)
33. Vivekananthan, V., Khandelwal, G., Alluri, N.R., Kim, S.J.: E-skin for futuristic nanosensor technology for the healthcare system. In: Nanosensors for Futuristic Smart and Intelligent Healthcare Systems, pp. 133–157. CRC Press (2022)
34. Ramasamy, M., Varadan, V.K.: Wireless nanosensor system for diagnosis of sleep disorders. In: Nanosensors, Biosensors, and Info-Tech Sensors and Systems 2016, vol 9802, pp. 200–205. SPIE (2016)
35. Pradhan, M.R., Mago, B., Ateeq, K.: A classification-based sensor data processing method for the internet of things assimilated wearable sensor technology. Clust. Comput. **26**(1), 807–822 (2023)
36. Tang, W.H., Ho, W.H., Chen, Y.J.: Data assimilation and multisource decision-making in systems biology based on unobtrusive internet-of-things devices. Biomed. Eng. Online **17**(2), 1–13 (2018)
37. Manogaran, G., Alazab, M., Song, H., Kumar, N.: CDP-UA: Cognitive data processing method wearable sensor data uncertainty analysis in the internet of things assisted smart medical healthcare systems. IEEE J. Biomed. Health Inform. **25**(10), 3691–3699 (2021)
38. Hassanalieragh, M., Page, A., Soyata, T., Sharma, G., Aktas, M., Mateos, G., et al.: Health monitoring and management using internet-of-things (IoT) sensing with cloud-based processing: opportunities and challenges. In: 2015 IEEE International Conference on Services Computing, pp. 285–292. IEEE.
39. Suma, D.V.: Wearable IoT based distributed framework for ubiquitous computing. J. Ubiq. Comput. Commun. Technol **3**(1), 23–32 (2021)
40. Jenkins, T.: Wearable medical sensor devices, machine and deep learning algorithms, and internet of things-based healthcare systems in COVID-19 patient screening, diagnosis, monitoring, and treatment. Am. J. Med. Res. **9**(1), 49–64 (2022)

41. Malik, A., Kumar, A.: Assimilation of blockchain with Internet of Things (IoT) with possible issues and solutions for better connectivity and proper security. In: New Trends and Applications in Internet of Things (IoT) and Big Data Analytics, pp. 187–207. Springer International Publishing, Cham (2022)
42. Byerly, K., Vagner, L., Grecu, I., Grecu, G., Lăzăroiu, G.: Real-time big data processing and wearable internet of medical things sensor devices for health monitoring. Am. J. Med. Res. **6**(2), 67–72 (2019)
43. Kumar, P.M., Khan, L.U., Hong, C.S.: Affirmative fusion process for improving wearable sensor data availability in artificial intelligence of medical things. IEEE Sens. J. (2022)
44. Han, X.: A novel assimilated navigation model based on advanced optical systems (AOS), internet of things (IoT) and artificial intelligence (AI). Opt. Quant. Electron. **55**(7), 655 (2023)
45. Dash, D., Farooq, R., Panda, J.S., Sandhyavani, K.V. Internet of things (IoT): the new paradigm of HRM and skill development in the fourth industrial revolution (Industry 4.0). IUP J. Inf. Technol. **15**(4) (2019)
46. Goethals, I.: Real-time and remote health monitoring internet of things-based systems: digital therapeutics, wearable and implantable medical devices, and body sensor networks. Am. J. Med. Res. **6**(2), 43–48 (2019)
47. Banerjee, A., Chakraborty, C., Rathi, Sr, M.: Medical imaging, artificial intelligence, internet of things, wearable devices in terahertz healthcare technologies. In: Terahertz Biomedical and Healthcare Technologies, pp. 145–165. Elsevier (2020)
48. Chen, E.T.: The internet of things: opportunities, issues, and challenges. In: The Internet of Things in the Modern Business Environment, pp. 167–187. IGI Global (2017)
49. Barnes, R., Zvarikova, K.: Artificial intelligence-enabled wearable medical devices, clinical and diagnostic decision support systems, and internet of things-based healthcare applications in COVID-19 prevention, screening, and treatment. Am. J. Med. Res. **8**(2), 9–22 (2021)
50. Desai, D., Shende, P.: Integration of internet of things with quantum dots: a state-of-the-art of medicine. Curr. Pharm. Des. **27**(17), 2068–2075 (2021)
51. Albalawi, U., Joshi, S.: Secure and trusted telemedicine in internet of things IoT. In: 2018 IEEE 4th World Forum on Internet of Things (WF-IoT), pp. 30–34. IEEE (2018)
52. Hurley, D., Popescu, G.H.: Medical big data and wearable internet of things healthcare systems in remotely monitoring and caring for confirmed or suspected COVID-19 patients. Am. J. Med. Res. **8**(2), 78–90 (2021)
53. Singh, D., Divan, M., Singh, M.: Internet of things for smart community solutions. Sensors **22**(2), 640 (2022)
54. Tsang, Y.P., Choy, K.L., Wu, C.H., Ho, G.T., Lam, C.H., Koo, P.S.: An internet of things (IoT)-based risk monitoring system for managing cold supply chain risks. Ind. Manag. Data Syst. **118**(7), 1432–1462 (2018)
55. Goel, S.S., Goel, A., Kumar, M., Moltó, G.: A review of internet of things: qualifying technologies and boundless horizon. J. Reliab. Intell. Environ. **7**, 23–33 (2021)
56. Kumar, P.M., Khan, L.U., Hong, C.S.: Notice of violation of IEEE publication principles: affirmative fusion process for improving wearable sensor data availability in artificial intelligence of medical things. IEEE Sens. J.
57. Huang, J., Wu, X., Huang, W., Wu, X., Wang, S.: Internet of things in health management systems: a review. Int. J. Commun. Syst. **34**(4), e4683 (2021)
58. Malik, P.K., Sharma, R., Singh, R., Gehlot, A., Satapathy, S.C., Alnumay, W.S., et al.: Industrial internet of things and its applications in industry 4.0: state of the art. Comput. Commun. **166**, 125–139 (2021)
59. Singh, B.: Affordability of medicines, public health and TRIPS regime: a comparative analysis. Indian J. Health Med. Law **2**(1), 1–7 (2019)
60. Kumar, P.M., Hong, C.S., Afghah, F., Manogaran, G., Yu, K., Hua, Q., Gao, J.: Clouds proportionate medical data stream analytics for internet of things-based healthcare systems. IEEE J. Biomed. Health Inf. **26**(3), 973–982 (2021)
61. Kavre, M., Gadekar, A., Gadhade, Y.: Internet of things (IoT): a survey. In: 2019 IEEE Pune Section International Conference (PuneCon), pp. 1–6. IEEE (2019)

62. Abraham, S.C.: Internet of things (IoT) with cloud computing and machine-to-machine (M2M) communication. IJETST **3**, 4654–4661 (2016)
63. Xu, Z.: Blockchain-design-for-a-secure-pharmaceutical supply-chain (2024)
64. Zhu, H., Wu, C.K., Koo, C.H., Tsang, Y.T., Liu, Y., Chi, H.R., Tsang, K.F.: Smart healthcare in the era of internet-of-things. IEEE Consum. Electron. Mag. **8**(5), 26–30 (2019)
65. Bhatt, C., Dey, N., Ashour, A.S. (eds.): Internet of things and big data technologies for next generation healthcare (2017)
66. Dimitrov, D.V.: Medical internet of things and big data in healthcare. Healthcare Inf. Res. **22**(3), 156–163 (2016)
67. Javaid, M., Khan, I.H.: Internet of things (IoT) enabled healthcare helps to take the challenges of COVID-19 pandemic. J. Oral Biol. Craniof. Res. **11**(2), 209–214 (2021)
68. Mukati, N., Namdev, N., Dilip, R., Hemalatha, N., Dhiman, V., Sahu, B.: Healthcare assistance to COVID-19 patient using internet of things (IoT) enabled technologies. Mater. Today Proc. **80**, 3777–3781 (2023)
69. West, D.M.: How 5G technology enables the health internet of things. Brook. Center Technol. Innov. **3**(1), 20 (2016)
70. Tien, J.M.: Internet of things, real-time decision making, and artificial intelligence. Ann. Data Sci. **4**, 149–178 (2017)
71. Zhang, Z., Wen, F., Sun, Z., Guo, X., He, T., Lee, C.: Artificial intelligence-enabled sensing technologies in the 5G/internet of things era: from virtual reality/augmented reality to the digital twin. Adv. Intell. Syst. **4**(7), 2100228 (2022)
72. Jagadeeswari, V., Subramaniyaswamy, V., Logesh, R., Vijayakumar, V.: A study on medical internet of things and big data in personalized healthcare system. Health inf. Sci. Syst. **6**, 1–20 (2018)
73. Rejeb, A., Rejeb, K., Treiblmaier, H., Appolloni, A., Alghamdi, S., Alhasawi, Y., Iranmanesh, M.: The internet of things (IoT) in healthcare: taking stock and moving forward. Internet Things **100721** (2023)
74. Aceto, G., Persico, V., Pescapé, A.: Industry 4.0 and health: internet of things, big data, and cloud computing for healthcare 4.0. J. Indus. Inf. Integr. **18**, 100129 (2020)
75. Asif-Ur-Rahman, M., Afsana, F., Mahmud, M., Kaiser, M.S., Ahmed, M.R., Kaiwartya, O., James-Taylor, A.: Toward a heterogeneous mist, fog, and cloud-based framework for the internet of healthcare things. IEEE Internet Things J. **6**(3), 4049–4062 (2018)
76. Metcalf, D., Milliard, S.T., Gomez, M., Schwartz, M.: Wearables and the internet of things for health: wearable, interconnected devices promise more efficient and comprehensive health care. IEEE Pulse **7**(5), 35–39 (2016)
77. Kang, M., Park, E., Cho, B.H., Lee, K.S.: Recent patient health monitoring platforms incorporating internet of things-enabled smart devices. Int. Neurourol. J. **22**(Suppl 2), S76 (2018)
78. Manogaran, G., Thota, C., Lopez, D., Vijayakumar, V., Abbas, K.M., Sundarsekar, R.: Big data knowledge system in healthcare. In: Internet of Things and Big Data Technologies for Next Generation Healthcare, pp. 133–157 (2017)
79. Singh, B.: Blockchain technology in renovating healthcare: legal and future perspectives. In: Revolutionizing Healthcare Through Artificial Intelligence and Internet of Things Applications, pp. 177–186. IGI Global (2023)
80. Zhan, K.: Sports and health big data system based on 5G network and internet of things system. Microprocess. Microsyst. **80**, 103363 (2021)
81. Qi, J., Yang, P., Min, G., Amft, O., Dong, F., Xu, L.: Advanced internet of things for personalised healthcare systems: a survey. Pervas. Mob. Comput. **41**, 132–149 (2017)
82. Firouzi, F., Rahmani, A.M., Mankodiya, K., Badaroglu, M., Merrett, G.V., Wong, P., Farahani, B.: Internet-of-things and big data for smarter healthcare: from device to architecture, applications and analytics. Futur. Gener. Comput. Syst. **78**, 583–586 (2018)
83. Al Hayani, B., Ilhan, H.: Image transmission over decode and forward based cooperative wireless multimedia sensor networks for Rayleigh fading channels in medical internet of things (MIoT) for remote health-care and health communication monitoring. J. Med. Imaging Health Inf. **10**(1), 160–168 (2020)

84. Singh, B.: COVID-19 pandemic and public healthcare: endless downward spiral or solution via rapid legal and health services implementation with patient monitoring program. Justice Law Bull. **1**(1), 1–7 (2022)
85. Gupta, P., Agrawal, D., Chhabra, J., Dhir, P.K.: IoT based smart healthcare kit. In: 2016 International Conference on Computational Techniques in Information and Communication Technologies (ICCTICT), pp 237–242. IEEE (2016)
86. Rahmani, A.M., Gia, T.N., Negash, B., Anzanpour, A., Azimi, I., Jiang, M., Liljeberg, P.: Exploiting smart e-health gateways at the edge of healthcare internet-of-things: a fog computing approach. Futur. Gener. Comput. Syst. **78**, 641–658 (2018)
87. Wang, W., Kumar, N., Chen, J., Gong, Z., Kong, X., Wei, W., Gao, H.: Realizing the potential of the internet of things for smart tourism with 5G and AI. IEEE Netw. **34**(6), 295–301 (2020)
88. He, L., Eastburn, M., Smirk, J., Zhao, H.: Smart chemical sensor and biosensor networks for healthcare 4.0. Sensors **23**(12), 5754 (2023)
89. Noah, N.M., Ndangili, P.M.: Nanosensor arrays. Nanosens. Futuris. Smart Intell. Healthcare Syst. **350** (2022)
90. Asorey-Cacheda, R., Correia, L.M., Garcia-Pardo, C., Wojcik, K., Turbic, K., Kulakowski, P.: Bridging nano and body area networks: a full architecture for cardiovascular health applications. IEEE Internet Things J. **10**(5), 4307–4323 (2022)
91. Uddin, M.H., Hossain, M.N., Ur Rahman, A.: A routing protocol for cancer cell detection using wireless nano-sensors network (WNSN). In: Proceedings of International Conference on Trends in Computational and Cognitive Engineering: Proceedings of TCCE 2020, pp. 569–578. Springer Singapore (2021)
92. Mbunge, E., Muchemwa, B., Batani, J.: Sensors and healthcare 5.0: transformative shift in virtual care through emerging digital health technologies. Global Health J. **5**(4), 169–177 (2021)
93. Varadan, V.K., Kumar, P.S., Oh, S., Kwon, H., Rai, P., Banerjee, N., Harbaugh, R.E.: E-nanoflex sensor system: smartphone-based roaming health monitor (2011)
94. Canovas-Carrasco, S., Asorey-Cacheda, R., Garcia-Sanchez, A.J., Garcia-Haro, J., Wojcik, K., Kulakowski, P.: Understanding the applicability of terahertz flow-guided nano-networks for medical applications. IEEE Access **8**, 214224–214239 (2020)
95. Javaid, S., Wu, Z., Hamid, Z., Zeadally, S., Fahim, H.: Temperature-aware routing protocol for intrabody nanonetworks. J. Netw. Comput. Appl. **183**, 103057 (2021)
96. Jin, H., Yu, J., Lin, S., Gao, S., Yang, H., Haick, H., et al.: Nanosensor-based flexible electronic assisted with light fidelity communicating technology for volatolomics-based telemedicine. ACS Nano **14**(11), 15517–15532 (2020)
97. Javaid, S., Zeadally, S., Fahim, H., He, B.: Medical sensors and their integration in wireless body area networks for pervasive healthcare delivery: a review. IEEE Sens. J. **22**(5), 3860–3877 (2022)
98. Rolfe, P.: Impact of micro and nano sensors in biomedical measurement. Key Eng. Mater. **437**, 299–303 (2010)
99. Jabeen, T., Jabeen, I., Ashraf, H., Ullah, A., Jhanjhi, N.Z., Ghoniem, R.M., Ray, S.K.: Smart Wireless Sensor Technology for Healthcare Monitoring System using Cognitive Radio Networks (2023)
100. Mohammad, H., Shubair, R.M.: Nanoscale communication: state-of-art and recent advances (2019). arXiv preprint arXiv:1905.07722
101. Yang, J., Carey IV, P., Ren, F., Lobo, B.C., Gebhard, M., Leon, M.E., et al.: Nanosensor networks for health-care applications. In: Nanosensors for Smart Cities, pp. 405–417. Elsevier (2020)
102. Adam, T., Gopinath, S.C.: Nanosensors: recent perspectives on attainments and future promise of downstream applications. Process Biochem. **117**, 153–173 (2022)
103. Varadan, V.K., Chen, L., Xie, J.: Nanomedicine: design and applications of magnetic nanomaterials, nanosensors and nanosystems. Wiley (2008)
104. Saylan, Y., Akgönüllü, S., Özgür, E., Denizli, A.: Nanosensors for smartphone-enabled sensing devices. In: Nanotechnology-Based Smart Remote Sensing Networks for Disaster Prevention, pp. 85–104. Elsevier (2022)

105. Al-Rawahi, M.N., Sharma, T., Palanisamy, P.: Internet of nanothings: challenges & opportunities. In: 2018 Majan International Conference (MIC), pp. 1–5. IEEE (2018)
106. Angelov, G.V., Nikolakov, D.P., Ruskova, I.N., Gieva, E.E., Spasova, M.L.: Healthcare sensing and monitoring. In: Enhanced Living Environments: Algorithms, Architectures, Platforms, and Systems, pp. 226–262. Springer International Publishing, Cham (2019)
107. Khazaei, M., Hosseini, M.S., Haghighi, A.M., Misaghi, M.: Nanosensors and their applications in early diagnosis of cancer. Sens. Bio-Sens. Res. 100569 (2023)
108. Kim, S.: Healthcare revolution: the power of nanosensing. Int. Stud. J. Med. **2**(2–3), 55–61 (2016)
109. Yang, L., Webster, T.J.: Monitoring tissue healing through nanosensors. In: Nanotechnology Enabled In Situ Sensors for Monitoring Health, pp. 41–59 (2011)
110. Wang, L., Lou, Z., Jiang, K., Shen, G.: Bio-multifunctional smart wearable sensors for medical devices. Adv. Intell. Syst. **1**(5), 1900040 (2019)
111. Haroun, A., Le, X., Gao, S., Dong, B., He, T., Zhang, Z., et al.: Progress in micro/nano sensors and nanoenergy for future AIoT-based smart home applications. Nano Exp. **2**(2), 022005 (2021)
112. Akyildiz, I.F., Ghovanloo, M., Guler, U., Ozkaya-Ahmadov, T., Sarioglu, A.F., Unluturk, B.D.: PANACEA: an internet of bio-nanothings application for early detection and mitigation of infectious diseases. IEEE Access **8**, 140512–140523 (2020)
113. Prabhu, R.S., Ananthi, D.S., Rajasoundarya, S., Janakan, R., Priyanka, R.: Internet of Nanothings (IoNT)—a concise review of its healthcare applications and future scope in pandemics (2021)
114. Das, P., Farihah, R.: Fundamental Application of Internet of Nano Things. Doctoral dissertation, Brac University (2019)
115. Chelliah, R., Khan, I., Wei, S., Madar, I.H., Sultan, G., Daliri, E.B.M., et al.: Intelligent packaging systems: food quality and intelligent medicine box based on nano-sensors. In: Smart Nanomaterials in Biomedical Applications, pp. 555–587. Springer International Publishing, Cham (2022)
116. Naranjo-Hernández, D., Reina-Tosina, J., Roa, L.M.: Special issue "Body sensors networks for e-health applications." Sensors **20**(14), 3944 (2020)
117. Olatinwo, D.D., Abu-Mahfouz, A., Hancke, G.: A survey on LPWAN technologies in WBAN for remote health-care monitoring. Sensors **19**(23), 5268 (2019)
118. Solanki, M.S., Nayak, M.M.: Survey on internet of nano things (iont). Technology **11**(10), 275–280 (2020)
119. De Pretis, F., van Gils, M., Forsberg, M.M.: A smart hospital-driven approach to precision pharmacovigilance. Trends Pharmacol. Sci. (2022)
120. Ayaz, M., Ammad-Uddin, M., Baig, I.: Wireless sensor's civil applications, prototypes, and future integration possibilities: a review. IEEE Sens. J. **18**(1), 4–30 (2017)
121. Sagar, A.K., Banda, L., Sahana, S., Singh, K., Singh, B.K.: Optimizing quality of service for sensor enabled Internet of healthcare systems. Neurosci. Inf. **1**(3), 100010 (2021)
122. Jabeen, T., Ashraf, H., Ullah, A.: A survey on healthcare data security in wireless body area networks. J. Amb. Intell. Hum. Comput. 1–14 (2021)
123. Karatas, M., Eriskin, L., Deveci, M., Pamucar, D., Garg, H.: Big data for healthcare industry 4.0: applications, challenges and future perspectives. Exp. Syst. Appl. **200**, 116912 (2022)
124. Fu, Q., Wu, Z., Xu, F., Li, X., Yao, C., Xu, M., et al.: A portable smart phone-based plasmonic nanosensor readout platform that measures transmitted light intensities of nanosubstrates using an ambient light sensor. Lab Chip **16**(10), 1927–1933 (2016)
125. Fouad, H., Hashem, M., Youssef, A.E.: Retracted article: a nano-biosensors model with optimized bio-cyber communication system based on Internet of Bio-Nano Things for thrombosis prediction. J. Nanopart. Res. **22**, 1–17 (2020)
126. Gulec, O.: Distributed routing and self-balancing algorithm for Medical IoNT. Simul. Model. Pract. Theory **129**, 102833 (2023)
127. Muthukaruppankaruppiah, S., Nagalingam, S.R., Murugasen, P., Nandaamarnath, R.: Human fatty liver monitoring using nano sensor and IoMT. Intell. Autom. Soft Comput. **35**(2) (2023)

128. Jabeen, T., Jabeen, I., Ashraf, H., Jhanjhi, N.Z., Yassine, A., Hossain, M.S.: An intelligent healthcare system using IoT in wireless sensor network. Sensors **23**(11), 5055 (2023)
129. Ji, Y., Du, W., Pan, S.L.: Vendor capabilities development in blockchain sourcing: a parallel play approach. J. Inf. Technol. 02683962231226399 (2024)
130. Balasubramani, S., Dhanalakshmi, R., Kavisankar, L., Ramesh, K., Saritha, S., Pandey, D.: Revolutionizing supply chain with machine learning and blockchain integration. In: Utilization of AI Technology in Supply Chain Management, pp. 113–125. IGI Global (2024)

Chapter 6
Blockchain Technology-Enabled Baggage Management at the Airport

R. Vedapradha and Deepika Joshi

Abstract The air transportation industry faces a persistent challenge in ensuring efficient and reliable baggage handling, which significantly impacts customer satisfaction and airport operations. This article presents a conceptual review of blockchain-based baggage management in airport operations, with the goal of streamlining the existing process. Using the properties of blockchain, which is decentralized, transparent, and non-erasable, the presented system strives to enhance the tracing, classification, and transportation of baggage from passengers. The introduction of blockchain technology is anticipated to reduce mishandling, wear & tear, and provide accountability to all players in the baggage handling chain. This review presents a systematic review of the existing baggage management issues at airports, the possible applications of blockchain technology, and a conceptual model of its implementation. The work helps enable the implementation of a more optimal, safe, and customer-oriented baggage management system, which, during the final stage, will improve the experience of airport passengers.

Keywords Blockchain · Airport · Baggage management · Customer operations · Conceptual review

6.1 Opening the Dialogue

"You didn't make it to the destination, but your luggage did". Consider a scenario in which you arrive at your destination without your bags, particularly if you are travelling by air. Despite the technological advancements in the aviation sector that are intended to improve productivity and the customer experience, mishandled passenger baggage is a major worry for airlines, airports, ground handlers, and passengers, and

R. Vedapradha
St. Joseph's College of Commerce, Bengaluru, India

D. Joshi (✉)
St. Joseph's Institute of Management, Bengaluru, India
e-mail: deepika@sjim.edu.in

P. M. Jeyanthi (ed.), *Distributed Ledger Technology in Communication*, Signals and Communication Technology, https://doi.org/10.1007/978-3-032-07682-3_6

it gets worse as demand rises. Baggage that has been lost, damaged, delayed, or stolen is considered mishandled.

Heathrow Airport Chaos (2022): Staffing shortages and a rise in passenger volume caused major baggage handling difficulties at London's Heathrow Airport. Media coverage and general annoyance resulted from thousands of travellers delayed or misplaced bags. Schiphol Airport in Amsterdam (2022): During the summer traffic season, Schiphol Airport in Amsterdam experienced similar problems with improper handling of baggage. A spike in passengers and a shortage of ground handling personnel caused the airport to suffer, leading to multiple claims of delayed and misplaced luggage. Incident at British Airways: In 2017, a worldwide IT breakdown at British Airways led to hundreds of flights being cancelled and extensive baggage disruption. The airline struggled to organize and deliver the mishandled bags, leaving thousands of people without their belongings for days. Flights and baggage handling were impacted by a system-wide breakdown that occurred in 2016 due to a power outage at Delta Air Lines' headquarters. Several travellers encountered considerable delays in getting their bags while the airline attempted to resume regular operations.

The global mishandled baggage rate has fluctuated over the years, with the rate in 2019 being roughly 5.57 bags per thousand travellers, according to the 2023 Baggage IT Insights study by SITA, a prominent air transport IT and communications provider. 2020: The rate decreased to 3.5 bags per thousand travellers as a result of the COVID-19 epidemic and a sharp decline in air travel. 2022: When travel resumed, the rate rose to 4.35 bags per thousand passengers, suggesting that mishandling problems had returned as airlines adapted to the sharp rise in passengers.

Technological implementation in the business has proved to be one of the most competitive advantages to various business firms that have the ability to transform the operations at a rapid pace. Automation in the business helps in improving operational efficiency, cost reduction, and streamlines the highly valued business processes. Some of the advanced technologies that are applied in both manufacturing and service sectors involve cloud computing, Artificial Intelligence (AI), blockchain technology, Internet of Things (IoT). Cloud computing assists business firms to access resources and software over the internet without the need for hardware infrastructure in the form of servers. This ensures scalability and flexibility with improved cost leading to better solutions [27]. AI has revolutionized the business operations where the repetitive and mundane tasks are automated, identifies the patterns, predicts models, analyses trends that help data-driven decision-making among the stakeholders [8, 10, 36]. Blockchain technology is a boon to the various industries due to its decentralized ledger that is more secure, transparent, and tamper-proof, which promotes better accuracy and eliminates the dependence on intermediaries [5, 10, 21]. IoT is connecting devices and sensors to the internet, enabling real-time data collection in industries like manufacturing, healthcare, and transportation by improving efficiency, reducing costs, and improving safety.

The presented research work utilizes the hermeneutic method to draw deeper understanding and insights into the use of blockchain to optimize baggage handling. This will steadily develop the literature upon a comprehensive understanding of the application of blockchain technology in the baggage management of passenger airlines. Numerous studies have investigated blockchain's possibilities in a range of sectors, including logistics and supply chain management as well [21, 44, 45, 66]. But a research work specific to luggage management solutions still exists as a research gap. The novelty of the research work lies in its linkage among supply chain management, cybersecurity, and information systems to study the possible advantages and difficulties of implementing blockchain in baggage management system.

6.2 Knowledge Nexus

Baggage handling mechanism is one of the most critical processes within the airline industry that requires excellent coordination between ground staff, cabin crew, and the support staff who are loading the luggage onto the respective aircraft [30, 47]. A thorough checking of this operational process within the system [2, 4, 46] must be integrated within the journey time of the passengers with the help of real-time data generated using advanced technologies that facilitate monitoring and eliminating the loss of luggage [6, 27]. The destination-coded vehicle (DCV) role is to quickly transport luggage to an off-ramp at the gate [41]. A miniature simulation of a baggage handling system that incorporates all luggage-related components, developed consisting of arrival of passengers at check-in lines, baggage check-in, security screening, sorting, transportation to the aircraft, and loading [24, 32, 48]. Similarly, the development of a platform that models baggage handling as an integrated unit and the expansion and application of a vehicle traffic simulator program to the baggage handling issue [7, 16].

Baggage handling is the process of transporting passenger luggage from a check-in counter at a departure airport, onto a plane cargo hold and then to a collection point at an arrival airport. There is constant demand to improve the baggage handling procedure since passengers view it as a significant contributor to quality in airport and airline management systems [9]. This is a reason for growing competitiveness in air transportation sector. The majority of domestic and international hub airports handle a sizable amount of transfer luggage, which creates a sorting issue. Additionally, the sheer amount of handled baggage increases the chance of traffic jams, incorrect routing, and control issues. Additionally, the physical size of the airport frequently significantly lengthens travel times to and from the airplane, leaving less time for the actual handling and sorting of baggage [12,30].

With time, the airport's capacity has been exceeded by the number of passengers it was designed to handle. The baggage handling system, which consists of a series of conveyors that transfer checked baggage to the baggage unloading zones or the buffer zones, is a crucial piece of airport infrastructure. Simulator models are frequently used to assess system performance and evaluate options. At airports, a System Modelling Language (SysML) are implemented to model the luggage handling system, which offers a formal representation of the system and can improve system validation [32]. Worldwide adoption of RFID-based baggage handling systems is very widespread.

As a result, there is a successfully declining trend in the amount of mishandled luggage.

Airport baggage handling transport system (BHTS) designs must be optimized with regard to the minimization of total costs, energy usage, and carbon emissions in airport operations [14, 33]. Simulation-based baggage handling system merge analysis inquiries into the design and management of bottleneck merging in conveyor-based baggage handling systems, including the impact of the physical layout of the merge and the merging control algorithm [18, 20, 24, 33]. When such a physical configuration is in a favoured position, input variation has no impact on bag throughput performance. Bag throughput is affected by the layout of the merging. Other material handling systems, such as those used in manufacturing and warehousing, may be able to benefit from these findings [15, 17, 24].

Utilizing the Lean Six Sigma technique, BHS may continually optimize the flow of baggage by determining the root causes of mishandled baggage and developing solutions to improve its performance. Low system reliability and the large volume of bags passing through manual encoding stations were the two most significant issues, and these issues were addressed using workable and affordable solutions [4, 19]. The use of RFID tags in the air baggage handling application increases the capacity for luggage tracking, dispatching, and transportation in order to increase management effectiveness and user happiness. Based on the RFID networks, a distributed application for tracking airline baggage has been created. An experiment conducted in Beijing Capital International Airport resulted in the rate of accurate recognition and keeping an eye on the condition of the luggage in real time. The financial gains made possible by the widespread use of RFID in the baggage handling system are also described [22, 23, 25, 67].

The minimal queue length rule is used by passengers to determine which check-in to use. Passengers are required to leave their bags when registering. Population density, societal wealth growth, and the health of industry are a few variables that affect the number of passengers. Discrete-event simulation module in a contemporary simulation system. Possibility of developing a graphical representation of the developed model in 2D and 3D. The experiments continued to use the constructed basis model [28, 30, 48]. The automation of airport baggage handling now relies on centralized control systems and traditional automation programming methods. For the growing demand for systems that are reconfigurable, fault-tolerant, and simple to maintain, these legacy automation technologies are becoming more and more constrained in this and other sectors of manufacturing and materials handling. A developing architectural framework called function blocks is used to create reusable parts for distributed industrial automation systems [11.

Several designs that incorporate function blocks were implemented for multiagent and holonic control systems. The effectiveness of the agent-based control system was demonstrated utilizing a multiagent control technique for a baggage handling system (BHS) employing IEC 61499 function blocks, and a utility is offered for real-time observation of these systems. The automation object technique is used in the design, which results in a function block component that represents a single conveyor section [29, 31, 37, 39, 40, 63]. Modern baggage handling systems convey bags automatically utilizing vehicles with destination codes (DCVs). These vehicles use a system of tracks to convey the bags at rapid speeds. Currently, routing strategies based on preferred paths are used to route the DCVs through the system.

Disruptive technologies are having a significant impact on the business models of airlines and are energizing the industry for not only short-term efficiency gains but also for accelerating differentiation and aiding in the achievement of long-term goals of improving customer experience and boosting profitability. They combine data, people, and procedures to effectively optimize resources and make decisions that will increase revenues [13].

Algorithms were used for the routing of some planned occurrences. In order to maximize the system's performance, efficient centralized, decentralized, and distributed forecasting algorithms were developed and tested [38, 49, 51, 52]. From the standpoint of disruptive innovation, the long-haul low-cost (LHLC) airline business model's effect on the aviation sector has been studied. The LHLC business model is here to stay and is destined to take over the incumbent network carriers' current share of the mass market. As a result, the long-haul air travel sector will experience major strategic changes; therefore, incumbents would be well to prepare their replies [3, 50, 53, 55, 59, 61, 62].

In baggage handling, often there are minimum of five stakeholders involved—passengers, airlines, airports, security, and handling agents. Each of these has a significant role and may have competing interests. It has always been a contentious issue with airlines in certain sectors that, due to payload restrictions at airports and on aircrafts baggage is left behind. Thus, baggage handling has historically been a difficult issue with airlines in specific sectors [67]. Baggage transfer manually from airport terminal to aircraft and vice versa is time-consuming, costly and labour-intensive operations. The end-to-end process of baggage loading and unloading decides the turnaround time for any aircraft and thus is a critical indicator of airline operational efficiency. In this, screening, sorting, and transportation activities are among few issues that have direct impact upon airlines responsiveness. Similarly, transfer of baggage on connecting flights, especially within an alliance and an international transfer, depends upon technology integration [34, 54, 55, 56, 57, 58, 60]. An understanding of operational challenges owing to baggage handling in air transportation.

Airports' efficient handling and transportation of passenger luggage and cargo shipments is made possible by baggage and cargo handling systems [42]. A baggage handling system (BHS), a type of conveyor system installed to transport checked baggage from check-in desk to the areas where bags can be loaded onto aircraft, comprises of a number of different processes and checks. It is designed to automatically count bags, weigh them, balance loads, scan and screen them for security, carry them through an airport's conveyor belt system, and read their information [67]. Here, the failure of carts leads to a process failure. Like, the issue of luggage cart failure in the high-speed baggage tunnel at Heathrow Terminal 5. The use of advanced blockchain-based technology can generate real-time data help to prevent luggage cart failure, condition monitoring and condition-based maintenance [11, 27]. Similarly, the use of smart technologies empowered in Indonesia's development of

Table 6.1 Proposed hypothesis

Symbols	Proposed hypothesis
H_1	Blockchain can reduce the rate of lost or mishandled baggage
H_2	Blockchain can improve the efficiency of baggage handling processes
H_3	Blockchain can enhance the security and privacy of passenger data

Source Author's own work

an RFID-based baggage handling system [9]. The use of blockchain technology in reducing failures and bringing process efficiency.

Along with three pillars of management, i.e., people, process, and technology, the proliferation of innovations like app-enabled tracking, Barcode Automatic Tag Readers (ATRs), RFID (Radio Frequency Identification) ATRs, and image capture with Artificial Intelligence (AI) computer vision technology has eased the airlines operations [1, 9, 26]. But the use of blockchain for process improvement in airlines industry specific to baggage handling is yet to be realized.

6.3 Tools of the Trade

Baggage handling system has three main functions: First, move bags from check-in area into the departure gate. Second, move bags from one gate to another during transfer. Thirdly, move bags from arrival gate to baggage claim area. Considering the challenges faced by the airline industry in handling their luggage, the following research questions are attempted to be answered:

RQ_1: How can blockchain technology improve baggage management efficiency and security in the airline industry?

RQ_2: How can smart contracts on blockchain automate and streamline baggage claim process in airline industry?

Based on the literature review, three hypotheses are proposed. The same is given in Table 6.1.

6.4 Blockchain Application in Tourism and Airlines

The automation of baggage handling will likely require a blockchain supply chain system to track luggage. Automated blockchain luggage management will ensure the safe arrival of luggage on the correct flight, reducing insurance and recovery costs. Passenger satisfaction will also be improved with reduced check-in times. Blockchain technology is an important tool for tracking luggage status and location, as well as passenger identifying information. The use of blockchain in air cargo will improve overall flight performance and speed up cargo delivery from point of origin to point of destination [44].

It is said that blockchain technology will make it easier and less complex to integrate luggage handling and management systems over the cloud. Using efficient and open-source blockchain technology, luggage data is processed and stored. Any passenger can use the developed framework to track their luggage in real time over the cloud from any location at any time. The use of blockchain technology over a public cloud infrastructure enables this [35]. The key effects of blockchain technology on operations management (OM), with a focus on supply chain management (SCM) decision-making processes from the standpoint of sustainable performance. It analyses data by reading and processing financial statements, non-financial reports, and the website of one key airport infrastructure to cover the airport industry from the standpoint of sustainable performance. In order to lessen operational fragmentation, inefficiency, and uncoordinated behaviour, it encourages collaboration between the major aviation sector actors and the air traffic controllers (ATCs) [21].

The current state of airworthiness management and the traceability of domestic aircraft equipment focused on supply chain traceability model for the airworthiness of aero plane equipment using blockchain and Internet of Things technology. The Internet of Things technology has become relatively developed through research on the state of the art in the civil aviation industry in Europe and the United States, but the application research of blockchain technology in the management of aircraft equipment is still in its infancy [64, 65, 68].

6.5 Theory Development

Blockchain can contribute to the development of more efficient and resilient supply chain networks in the airline industry. The application of blockchain in baggage management can advance the field of information systems and cybersecurity. Aviation authorities may need to develop regulatory frameworks to facilitate the adoption of blockchain technology. Industry standards and protocols may be required to ensure interoperability between different blockchain systems. Operational terminologies that are integrated are illustrated in Table 6.2.

Blockchain can improve the flow of information and physical assets throughout the baggage handling process by providing a transparent, secure, and efficient system for tracking and managing baggage from check-in to delivery. It allows for the assignment of a distinct digital identity to every item of luggage, which is then recorded on a distributed ledger. This enables real-time, unchangeable information regarding the location and status of the baggage at any time during the process to be accessed by all parties (airlines, baggage handlers, ground personnel, and passengers). Blockchain can guarantee precise and timely updates while tracking the movement of luggage through several stages, including check-in, security, loading, transfer, and delivery. Blockchain's decentralized structure minimizes errors and inconsistencies by guaranteeing that all parties involved have access to the same data. This technology

Table 6.2 Summary of operational terms

S. No.	Operational term	Operational definition
1	Baggage tag	A unique identifier for each piece of baggage
2	Internet of Things (IoT)	Network of interconnected devices that collect and exchange data
3	Smart contract	Self-executing contracts with the terms of the agreement directly written into code

Source Authors' Compilation

offers better security and can confidently help in handling process of passenger luggage as these are decentralized, immutable digital records without leaving scope for tampering with the transactions without the mutual consensus of all the participants in the network. It encourages smooth tracking of any luggage, if lost without delaying the process, thus assisting the crew of the airline industry and ensuring the safety of the baggage (Fig. 6.1).

Blockchain's decentralized nature and cryptographic techniques enhance security and privacy. It offers an unchangeable ledger in which all baggage-related information is decentralized and recorded, including movement logs, baggage tracking, and handovers. It doesn't give scope to hackers or fraudsters in manipulating the data pertaining to handling the passenger luggage, considering the dynamics of the challenges involved especially when there are chances of anti-social activities like carrying drugs, explosives, arms, etc. This traceable feature of this technology empowers all the stakeholders (passengers, baggage handlers, and airlines) with greater transparency by preventing data tampering, such as concealing delays or misdirecting bags.

Conventional baggage management system involves manual intervention that is dependent on centralized databases, leaving more scope to human errors, involvement of illegal and anti-social elements. The technology ensures the data privacy,

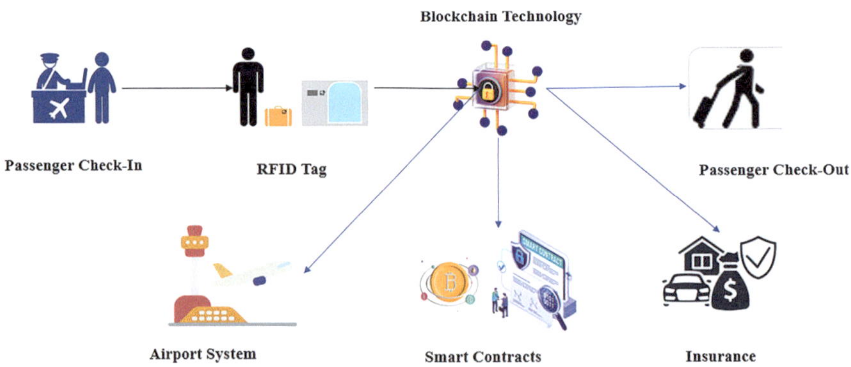

Fig. 6.1 Blockchain technology integrated baggage handling system. *Source* Authors' Work

preventing unauthorized access outside the blockchain network when managed through private or public domain. Blockchain protects data with cutting-edge encryption algorithms. Only authorized parties can access the encrypted data that is stored when a passenger's baggage is logged on the blockchain. The blockchain makes sure that the information can only be viewed or updated by those who have the right private keys. This lowers the possibility of identity theft and privacy violations while safeguarding private passenger information, including travel schedules and personal information, against unwanted access. To ensure privacy while maintaining tracking capabilities, blockchain may, for example, retain only the most basic information about the baggage itself, such as its weight and flight number, without disclosing personal information.

Blockchain protects data with cutting-edge encryption algorithms. Only authorized parties can access the encrypted data that is stored when a passenger's baggage is logged on the blockchain. The blockchain makes sure that the information can only be viewed or updated by those who have the right private keys. This lowers the possibility of identity theft and privacy violations while safeguarding private passenger information, including travel schedules and personal information, against unwanted access. To ensure privacy while maintaining tracking capabilities, blockchain may, for example, retain only the most basic information about the baggage itself, such as its weight and flight number, without disclosing personal information. Information in conventional baggage management systems may be manipulated or misused by staff members or service providers who have access to central databases because blockchain is decentralized, no one entity has complete control over the system, lowering this risk. Even if an insider (such as a luggage handler or an airline employee) tries to alter data, the consensus process of the blockchain makes sure that other network users will be able to identify and reject these attempts.

"Smart contracts" can be integrated into blockchain to automate specific steps in the luggage handling procedure eliminating human intervention. It can speed up the process of handling the baggage in terms of scanning and automatically recording the barcode that is attached to each luggage, which improves the efficiency. They help in easing the process by bringing all the parties within the network, where each party has access to track the luggage in real time. This also provides monitoring at the ground level and informs the concerned authority if there is any misplaced baggage. It also helps in rerouting the wrongly loading luggage to the concerned destination, along with informing the staff promoting easy baggage transfer based on pre-determined specific criteria when fulfilled by these self-executing contracts. This mitigates the fraudulent practices and loss of baggage as these are encoded within the blockchain technology that can't be manipulated without the mutual consensus of all the stakeholders in the network. It strengthens the cybersecurity measures by reducing the risk of cyberattacks, prevents data loss, encourages data privacy creating visible and safe records of the baggage movements. For instance, if a piece of luggage hasn't been loaded onto the appropriate flight, a smart contract might immediately sound an alert, guaranteeing that any mistakes are found quickly and that data is kept accurate and safe.

6.6 Future Road Map

The scalability and energy consumption of blockchain technology may pose challenges for large-scale implementation. Further research is needed to explore the economic and environmental impacts of blockchain-based baggage management. Future research can employ case studies, surveys, and simulations to evaluate the impact of blockchain on baggage management. Experimental studies can be conducted to test the feasibility and effectiveness of blockchain-based solutions.

6.7 Conclusion

The airline industry's baggage management problems may be resolved with the help of blockchain technology. Blockchain can improve the entire traveller's experience by increasing effectiveness, security, and transparency. For implementation to be successful, technical, financial, and regulatory aspects must be carefully taken into account. Blockchain has enormous potential for baggage management, but there are still technological obstacles and industry adoption hurdles to overcome. Blockchain helps cut down on inefficiencies that can result in compensation claims and operational delays, such as misrouted or lost luggage. Airlines and handling businesses can lower these operating expenses and increase the overall luggage process's speed and dependability with a more precise tracking system.

References

1. Ahmad, R.W., Salah, K., Jayaraman, R., Hasan, H.R., Yaqoob, I., Omar, M.: The role of blockchain technology in aviation industry. IEEE Aerosp. Electron. Syst. Mag. **36**(3), 4–15 (2021)
2. Akter, S., Michael, K., Uddin, M.R., McCarthy, G., Rahman, M.: Transforming business using digital innovations: the application of AI, blockchain, cloud and data analytics. Ann. Oper. Res. 1–33 (2020)
3. Albers, S., Daft, J., Stabenow, S., Rundshagen, V.: The long-haul low-cost airline business model: a disruptive innovation perspective. J. Air Transp. Manag. **89**, 101878 (2020)
4. Alsyouf, I., Kumar, U., Al-Ashi, L., Al-Hammadi, M.: Improving baggage flow in the baggage handling system at a UAE-based airline using lean Six Sigma tools. Qual. Eng. **30**(3), 432–452 (2018)
5. Angelis, J., Da Silva, E.R.: Blockchain adoption: a value driver perspective. Bus. Horiz. **62**(3), 307–314 (2019)
6. Antoniadis, I., Spinthiropoulos, K., Kontsas, S.: Blockchain applications in tourism and tourism marketing: a short review. Strat. Innov. Mark. Tour. 375–384 (2020)
7. Balasubramanian, S., Sethi, J.S., Ajayan, S., Paris, C.M.: An enabling framework for blockchain in tourism. Inf. Technol. Tour. 1–15 (2022)
8. Baralla, G., Pinna, A., Tonelli, R., Marchesi, M., Ibba, S.: Ensuring transparency and traceability of food local products: a blockchain application to a Smart Tourism Region. Concurr. Comput. Pract. Exp. **33**(1), e5857 (2021)

9. Baskoro, H., Prabowo, H., Trisetyarso, A., Hidayanto, A.N.: Design considerations of RFID based baggage handling system, a literature review. In: 2017 International Conference on Information Management and Technology (ICIMTech), pp. 210–214. IEEE (2017)
10. Bhardwaj, S., Kaushik, M.: Blockchain—technology to drive the future. In: Smart Computing and Informatics, pp. 263–271. Springer, Singapore (2017)
11. Black, G., Vyatkin, V.: Intelligent component-based automation of baggage handling systems with IEC 61499. IEEE Trans. Autom. Sci. Eng. **7**(2), 337–351 (2009)
12. Bucci, N., Luna, M., Viloria, A., Hernández García, J., Parody, A., Varela, N., Borrero López, L.A.: Factor analysis of the psychosocial risk assessment instrument. In: Conference on Data Mining and Big Data, pp. 149–158. Springer, Cham (2018)
13. Burger, R.G., Orr, B.: What business are airports really in? J. Airport Manage. **15**(2), 114–127 (2021)
14. Calvaresi, D., Leis, M., Dubovitskaya, A., Schegg, R., Schumacher, M.: Trust in tourism via blockchain technology: results from a systematic review. Inf. Commun. Technol. Tour. **2019**, 304–317 (2019)
15. Campbell, M.E., Keith, V.M., Gonlin, V., Carter-Sowell, A.R.: Is a picture worth a thousand words? An experiment comparing observer-based skin tone measures. Race Soc. Probl. **12**(3), 266–278 (2020)
16. Cavada, J.P., Cortés, C.E., Rey, P.A.: A simulation approach to modelling baggage handling systems at an international airport. Simul. Model. Pract. Theory **75**, 146–164 (2017)
17. Cheah, J.-H., Sarstedt, M., Ringle, C.M., Ramayah, T., Ting, H.: Convergent validity assessment of formatively measured constructs in PLS-SEM: on using single-item versus multi-item measures in redundancy analyses. Int. J. Contemp. Hosp. Manag. **30**(11), 3192–3210 (2018)
18. Chen, J.-S., Kerr, D., Chou, C.Y., Ang, C.: Business co-creation for service innovation in the hospitality and tourism industry. Int. J. Contemp. Hosp. Manag. **29**(6), 1522–1540 (2017)
19. Ciriello, R.F., Richter, A., Schwabe, G.: Digital innovation. Bus. Inf. Syst. Eng. **60**(6), 563–569 (2018)
20. Clohessy, T., Acton, T.: Investigating the influence of organizational factors on blockchain adoption: an innovation theory perspective. Ind. Manag. Data Syst. **119**(7), 1457–1491 (2019)
21. Di Vaio, A., Varriale, L.: Blockchain technology in supply chain management for sustainable performance: evidence from the airport industry. Int. J. Inf. Manage. **52**, 102014 (2020)
22. Hassi, A.: Empowering leadership and management innovation in the hospitality industry context: the mediating role of climate for creativity. Int. J. Contemp. Hosp. Manag. **31**(4), 1785–1800 (2019)
23. Irannezhad, E., Mahadevan, R.: Is blockchain tourism's new hope? J. Hosp. Tour. Technol. **12**(1), 85–96 (2021)
24. Johnstone, M., Creighton, D., Nahavandi, S.: Simulation-based baggage handling system merge analysis. Simul. Model. Pract. Theory **53**, 45–59 (2015)
25. Khalifa: Blockchain: The UPcoming revolution in the world of managment and finance. Futureuae (2018). Retrieved from: https://futureuae.com/media/Ehabpdf_d1f747f1-7ba7-4390-bd3f-918c5dbf6ead.pdf
26. Kim, G., Kim, J., Chae, J.: Balancing the baggage handling performance of a check-in area shared by multiple airlines. J. Air Transp. Manag. **58**, 31–49 (2017)
27. Koenig, F., Found, P.A., Kumar, M.: Condition monitoring for airport baggage handling in the era of industry 4.0. J. Qual. Maint. Eng. **25**(3), 435–451 (2019)
28. Koenig, F., Found, P.A., Kumar, M.: Innovative airport 4.0 condition-based maintenance system for baggage handling DCV systems. Int. J. Prod. Perform. Manag. **68**(3), 561–577 (2019)
29. Kwok, A.O., Koh, S.G.: Is blockchain technology a watershed for tourism development? Curr. Issue Tour. **22**(20), 2447–2452 (2019)
30. Lenior, O.N.M.: Airport baggage handling–where do human factors fit in the challenges that airports put on a baggage system? Work **41**(Supplement 1), 5899–5904 (2012)
31. Liang, T.P., Kohli, R., Huang, H.C., Li, Z.L.: What drives the adoption of the blockchain technology? A fit-viability perspective. J. Manag. Inf. Syst. **38**(2), 314–337 (2021)

32. Lin, J.T., Shih, P.H., Huang, E., Chiu, C.C.: Airport baggage handling system simulation modeling using SysML. In: 2015 International Conference on Industrial Engineering and Operations Management (IEOM), pp. 1–10. IEEE (2015)
33. Lodewijks, G., Cao, Y., Zhao, N., Zhang, H.: Reducing CO_2 emissions of an airport baggage handling transport system using a particle swarm optimization algorithm. IEEE Access **9**, 121894–121905 (2021)
34. Møller, K.L., Brauer, C., Mikkelsen, S., Loft, S., Simonsen, E.B., Koblauch, H., Thygesen, L.C.: Copenhagen Airport Cohort: air pollution, manual baggage handling and health. Bmj Open **7**(5), e012651 (2017)
35. Muruganantham, A., Joseph, B.: Smart airline baggage tracking and theft prevention with blockchain technology. Test Eng. Manag. **83**(3), 3436–3440 (2020)
36. Nam, K., Dutt, C.S., Chathoth, P., Khan, M.S.: Blockchain technology for smart city and smart tourism: latest trends and challenges. Asia Pac. J. Tour. Res. **26**(4), 454–468 (2021)
37. Nuryyev, G., Wang, Y.P., Achyldurdyyeva, J., Jaw, B.S., Yeh, Y.S., Lin, H.T., Wu, L.F.: Blockchain technology adoption behavior and sustainability of the business in tourism and hospitality SMEs: an empirical study. Sustainability **12**(3), 1256 (2020)
38. Önder, I., Gunter, U.: Blockchain: Is it the future for the tourism and hospitality industry? Tour. Econ. **28**(2), 291–299 (2022)
39. Ozdemir, A.I., Ar, I.M., Erol, I.: Assessment of blockchain applications in travel and tourism industry. Qual. Quant. 1–15 (2019)
40. Ozdemir, A.I., Ar, I.M., Erol, I.: Assessment of blockchain applications in travel and tourism industry. Qual. Quant. **54**(5), 1549–1563 (2020)
41. Pilkington, M.: Can blockchain technology help promote new tourism destinations? The example of medical tourism in Moldova. Ex. Med. Tour. Mold. (2017)
42. Rajapaksha, A., Jayasuriya, N.: Smart airport: a review on future of the airport operation. Glob. J. Manag. Bus. Res. (2020)
43. Rana, R.L., Adamashvili, N., Tricase, C.: The impact of blockchain technology adoption on tourism industry: a systematic literature review. Sustainability **14**(12), 7383 (2022)
44. Rashideh, W.: Blockchain technology framework: current and future perspectives for the tourism industry. Tour. Manage. **80**, 104125 (2020)
45. Rejeb, A., Karim, R.: Blockchain technology in tourism: applications and possibilities. World Sci. News **137**, 119–144 (2019)
46. Rezaei, J., Kothadiya, O., Tavasszy, L., Kroesen, M.: Quality assessment of airline baggage handling systems using SERVQUAL and BWM. Tour. Manage. **66**, 85–93 (2018)
47. Rijsenbrij, J.C., Ottjes, J.A.: New developments in airport baggage handling systems. Transp. Plan. Technol. **30**(4), 417–430 (2007)
48. Savrasovs, M., Medvedev, A., Sincova, E.: Riga airoport baggage handling system simulation. In: ECMS, pp. 384–390 (2009)
49. Sharma, M., Sehrawat, R., Daim, T., Shaygan, A.: Technology assessment: enabling blockchain in hospitality and tourism sectors. Technol. Forecast. Soc. Chang. **169**, 120810 (2021)
50. Si, P., Wang, F., Sun, E., Su, Y: BEI-TAB: enabling secure and distributed airport baggage tracking with hybrid blockchain-edge system. Wirel. Commun. Mob. Comput. (2021)
51. Swati, V., Prasad, A.S.: Application of blockchain technology in travel industry. In: International conference on circuits and systems in digital enterprise technology (ICCSDET), pp. 1–5. IEEE (2018)
52. Tarau, A.N., De Schutter, B., Hellendoorn, H.: Model-based control for route choice in automated baggage handling systems. IEEE Trans. Syst. Man Cybern. Part C (Applications and Reviews) **40**(3), 341–351 (2010)
53. Thees, H., Erschbamer, G., Pechlaner, H.: The application of blockchain in tourism: use cases in the tourism value system. Eur. J. Tour. Res. **26** (2020)
54. Toufaily, E., Zalan, T., Dhaou, S.B.: A framework of blockchain technology adoption: an investigation of challenges and expected value. Inf. Manage. **58**(3), 103444 (2021)
55. Treiblmaier, H.: Blockchain and tourism. Handbook of e-Tourism 1–21 (2020)

56. Treiblmaier, H., Önder, I.: The impact of blockchain on the tourism industry: a theory-based research framework. In: Business Transformation Through Blockchain, pp. 3–21. Palgrave Macmillan, Cham (2019)
57. Tumasjan, A., Beutel, T.: Blockchain-based decentralized business models in the sharing economy: a technology adoption perspective. In: Business Transformation Through Blockchain, pp. 77–120. Palgrave Macmillan, Cham (2019)
58. Tyan, I., Yagüe, M.I., Guevara-Plaza, A.: Blockchain technology's potential for sustainable tourism. In: Information and Communication Technologies in Tourism 2021, pp. 17–29. Springer, Cham (2021)
59. Uddin, M.A., Stranieri, A., Gondal, I., Balasubramanian, V.: A survey on the adoption of blockchain in IoT: challenges and solutions. Blockchain: Res. Appl. **2**(2), 100006 (2021)
60. Valeri, M.: Blockchain technology: adoption perspectives in tourism. In: Entrepreneurship and Organizational Change, pp. 27–35. Springer, Cham (2020)
61. Valeri, M., Baggio, R.: A critical reflection on the adoption of blockchain in tourism. Inf. Technol. Tour. **23**(2), 121–132 (2021)
62. Verma, M.: Application of hybrid blockchain and cryptocurrency in aviation. Int. J. Adv. Res., Ideas Innov. Technol. **4**(5), 819–824 (2018)
63. Vyatkin, V.: IEC 61499 function blocks for embedded and distributed control systems design. ISA (2007)
64. Wang, C., Li, B.: Research on traceability model of aircraft equipment based on blockchain technology. In: 2019 IEEE 1st International Conference on Civil Aviation Safety and Information Technology (ICCASIT), pp. 88–94. IEEE (2019)
65. Willie, P.: Can all sectors of the hospitality and tourism industry be influenced by the innovation of Blockchain technology? Worldw. Hosp. Tour. Themes (2019)
66. Yeasmin, S., Baig, A.: Unblocking the potential of blockchain. In: 2019 International Conference on Electrical and Computing Technologies and Applications (ICECTA), pp. 1–5. IEEE (2019)
67. Zhang, T., Ouyang, Y., He, Y.: Traceable air baggage handling system based on RFID tags in the airport. J. Theor. Appl. Electron. Commer. Res. **3**(1), 106–115 (2008)
68. Zhou, R., Wang, X., Shi, Y., Zhang, R., Zhang, L., Guo, H.: Measuring e-service quality and its importance to customer satisfaction and loyalty: an empirical study in a telecom setting. Electron. Commer. Res. **19**(3), 477–499 (2019)

Chapter 7
Optimizing Solar Energy Systems Through IoT, Metadata, and Blockchain Technologies

Gaurav Sharma and R. Oum Kumari

Abstract Climate change and global warming have become the greatest threat to the survival of lives on this planet. Rapidly increasing per capita consumption of energy has not only resulted in power crisis but also had affected the environment adversely as more than 60% of energy in India is derived from coal (Oum in Blockchain and Cryptocurrency: Management Systems and Technology Challenges, CRC, 2024). The continuous increase in the per capita consumption of electricity has forced the ministry of power to increase the installed capacity of power to overcome the problem of power deficit in India. The heavy dependence on coal has resulted in emission of 1165 million tons of carbon dioxide in the year 2022. Addressing climate change and energy shortage has been emphasized for last two decades, and renewable energy is a significant option to address energy and environmental challenges. Recent technologies like IoT, metadata, and blockchain technology can facilitate automation, decentralization, transparency, and audit trails in energy sector, and the present chapter highlights the importance of technologies in harnessing the solar power and overcome the power and environmental issues in long run.

Keywords Solar technologies · Climate change · Blockchain technology · Power crisis

7.1 Introduction

Energy consumption is considered to be an effective indicator of growth of an economy. Electricity is considered to be one of the important sources of energy, and the per capita electricity consumption has been growing an alarming rate in the last few decades. The per capita consumption of electricity in India was just 16 kilowatt

---

G. Sharma · R. Oum Kumari (✉)
Jaipuria Institute of Management, Jaipur, India
e-mail: oum.rathore@gmail.com

R. Oum Kumari
Jaipuria Institute of Management Jaipur, Jaipur, India

P. M. Jeyanthi (ed.), *Distributed Ledger Technology in Communication*, Signals and Communication Technology, https://doi.org/10.1007/978-3-032-07682-3_7

in 1947, and now it has increased to 1255 kilowatt in 2020–2024 (CEA report). The increase in the consumption of electricity accounting for 8000% is a positive indicator of growth and contributed to the increase in the GDP from $267.52 Billion in 1991 to around $4 Trillion in 2024 [World Bank 2024]. Electricity access undoubtedly has contributed to industrial growth, urbanization, and technological advancements since independence, but the efforts of ministry of power to meet the growing demand for power have imposed high penalty on the human life and the planet across the globe.

More than 70% of electricity is produced from fossil fuel specially coal. Coal is considered to be an immediate energy provider but accelerates emissions of harmful gases like carbon dioxide, sulfur dioxide, and nitrogen oxide. Coal combustion is also considered to be the largest emitter of CO_2 emission and accounting for more than 40% of total greenhouse gases which is leading to economic, environment, and social disruptions.

The following section of this chapter highlights the hidden cost of coal-generated electricity.

7.1.1 Climate Change

Electricity, being the biggest contributor to greenhouse gases, is considered as the prime cause of global warming and climate change [1]. CO_2 emissions from coal-fired plants in India are estimated to grow at the rate of 7% CAGR in the period 2005–2020. The electricity policy of 2005 announced free permitted captive generation and is also the first policy impetus that emphasized decarbonization. The continuous increase in the per capita consumption of electricity has forced the ministry of power to increase the installed capacity of power to overcome the problem of power deficit in India. The heavy dependence on coal has resulted in emission of 1165 million tons of carbon dioxide in the year 2022.

7.1.2 Health Hazards

Coal is considered as global energy source for many decades and has significant impact on the health of humans. Since coal is abundantly available on earth, it is extensively used for generating electricity across the globe. India alone accounts for 3–5% of total coal consumption per year (Seisminski and Administrator 2016). But the problem is with the quality of coal that has been used in thermal power stations which is very poor in quality [1] and results in significant amount of Combustion by Products (CCBs) [2]. The CCBs are considered to have a negative impact on the environment and human life (ACAA 2007, Sarkar et al. 2006). It is estimated that 217 million tons of ash is released in the air [3], and this fly ash is not managed appropriately, thus resulting in adverse health impact on life of humans and livestock. Out of 187 hazardous chemical compounds, 84 are released from thermal power plants (Environmental Protection Agency 2018). The hazardous chemical compounds are mobilized through soil and water channels, and they ultimately enter into the food channels harm the environment, and endanger the life of humans (Nordin et al., Oliveria et al.).

It is reported that the rapid increase in the cancer cases worldwide is linked to heavy metal poisoning (Zhou et al.), and coal power plants are one prime reason for this. India is the second largest consumer of energy that is of 3.4% of total power consumption and heavily relies on coal for its growing energy needs. The excessive use of coal has caused a significant threat not only to the environment but also to the life on this planet [Segun et al.].

So it is important to explore the potentials of renewables to overcome the problem of environmental degradation and power crisis to enable the economy to progress in the right direction.

7.2 Leveraging IoT, Metadata, and Blockchain for Renewable Energy Optimization

India is witnessing a massive transformation in the energy domain, wherein the contribution of renewable energy in the country's mix is increasing rapidly. By 2024, the total installed electricity generation capacity in the country has crossed 452.69 GW, wherein fossil fuels—coal, gas and oil—make up 56.8% and renewable sources 46.3%. The solar segment holds the leading position, with 90.76 GW installed capacity, followed by wind at 47.36 GW and hydropower at 46.92 GW. This leadership reflects the transition that India is making into low-carbon energy toward its sustainability and climate goals.

As one of the leading states for renewable energy in India, Rajasthan is a clear case of energy transitioning. Rajasthan has 29.98 GW of installed renewable capacity and, of course, leads with solar energy generation at 19.9 GW as of 2024. Other leaders are Gujarat with 29.52 GW and Tamil Nadu at 23.7 GW, which also rely on a mix of solar, wind, and hydroelectric energy resources. However, the massive desert spread of Rajasthan, along with the government schemes, makes it especially well-suited for the exploitation of the solar energy potential of the state. This is further supported by historical trends; the solar energy capacity in Rajasthan grew from a mere 1 GW in 2014 to almost 20 GW in ten years on its way, led by the National Solar Mission and investments in significant projects such as Bhadla Solar Park.

A high heterogeneity of the energy generation profile characterizes the top performers. Gujarat will join the well-established solar structure supplemented by wind farms along the coastline, and Tamil Nadu takes advantage of its windy conditions as one of the leaders in the wind energy generation. A strategy of Rajasthan, focusing primarily on solar, is supported with a policy structure and long history of commitment toward renewable sources of energy.

This section of the chapter deals with how the inclusion of cutting-edge technologies such as IoT, metadata, and blockchain further enhances India's and more specifically Rajasthan's solar potential, ensuring efficient generation, seamless trade, and greater rural electrification across the state.

7.2.1 IoT Applications in Solar Energy

The Internet of Things, or IoT, is a network of interconnected devices equipped with sensors, software, and communication technologies that allow them to collect and exchange data in real-time. This idea converts ordinary physical objects into a "smart" system that can monitor, analyze, and react to their surroundings autonomously. IoT applications integrate into many commercial devices, including smart thermostats, smart lighting systems, and voice-controlled assistance like Amazon Echo and Google Nest. These devices make life more effortless, reduce energy consumption, and provide unparalleled convenience. For example, IoT-powered refrigerators track inventory levels and alert them when items have reached their expiration dates; IoT-powered security networks allow homeowners to control their homes remotely using smart phone applications. For the renewable energy sector, functionality stretches much further beyond the wall of the home and truly changes the way solar energy is generated. IoT also can help to enhance the reliability and efficiency of solar plants because it allows for online tracking and analysis of data. These include.

(a) Real-Time Performance Monitoring: Sensors of the panel collect details on energy flow, temperature, and even environmental conditions to ensure that operations are at maximum levels.
(b) Predictive Maintenance: IoT systems analyze performance data and ensures that equipment downtime and repair costs minimal.
(c) Dynamic Solar Tracking by IoT-enabled solar trackers on panel orientation helps to maximize sunlight absorption.
(d) IoT-Based Energy Storage Management: IoT systems monitor and optimize battery storage solutions to ensure effective energy distribution during off-peak hours.xlsx.

There are a few companies in India such as DATOMS, Tata Power Solar, and Orb Energy that are applying IoT to optimize solar energy systems. These companies demonstrate how IoT can change the production of solar power by bringing efficiency, lowering costs, and helping India achieve its renewable energy goals. Rajasthan, with such large solar installations and an IoT-ready infrastructure, will greatly benefit from these advancements.pptx.

7.2.1.1 Significance of Radiation

IoT technology has improved the generation of solar farms by monitoring the panels in real-time and adjusting their orientation to suit the current solar irradiation. Sensors and actuators can automatically adjust the panel orientation by maximizing the amount of the sun absorbed by the photovoltaic panels. Dynamic tracking is useful in Rajasthan because high solar irradiation levels are intermittently interrupted by seasonal dust storms. For example, advanced IoT tracking technology has increased up to 30% the efficiency in Jodhpur of a solar farm. This makes an obvious impact that comes along with it in making energy production.pdf.

- IoT systems fundamentally transform the maintenance and operational culture of solar power plants by providing real-time data in their entirety. Sensors in the system continuously monitor energy production, panel temperature, and environmental factors surrounding the facility, which alerts anomalies before equipment failure. Such a proactive approach minimizes downtime and lowers maintenance cost. In Rajasthan, huge installations like Bhadla Solar Park have used IoT-enabled solutions, which have helped to reduce operational costs by 20% along with uninterrupted energy production.
- IoT-based mini-grids have immensely facilitated the electrification process in remote and isolated districts of Rajasthan. IoT gives flexibility to smart grids power supply according to demand. Technologies based on IoT can be implemented in such initiatives as Saubhagya for making the penetration of electrification to even inaccessible villages viable along with non-carbon-intensive energy supports toward socio-economic development.
- This section of the chapter highlights IoT fosters innovations in solar energy systems and the increase in efficiency at a lower cost, while helping to develop sustainable rural development of the nation.

7.2.2 Metadata in Solar Energy

Metadata is simply defined as "data about data" that provides one with necessary information on other datasets to help systems understand, categorize, and use them appropriately. In simple language, metadata is like a tag or description that explains what features of the data. For example, metadata related to a digital image denotes the time and location of the picture.

In the modern world where data are becoming more and more important to manage and analyze large datasets, metadata has become a very crucial factor. Beyond just descriptions, it has transformed many industries such as Health care, E-commerce, Transportation, Media, and Entertainment in the modern era.

Metadata has also revolutionized the solar power production within the renewable energy field by allowing good prediction, location selection, and performance improvement. Metadata serves as a base framework for incorporating a wide variety of information sources including ground sensors, meteorological satellites, and astronomical information systems. Raw data comprise temperature or solar irradiance measurements at a particular time. Metadata is contextual information that accompany the data, including the time and location of observation, the instrument used, and the calibration parameters. In this context, metadata is associated with trends in the brightness of the sun, which provides significant insight into seasonal as well as long-term changes. This link enables the predictive and adaptive capacity of solar energy systems, thus ensuring consistent output even amidst episodes of environmental variability.

The role of data providers is significant in the modern era. Agencies such as the Indian Meteorological Department (IMD) and private satellite companies provide high-resolution meteorological data, and energy companies input it into their forecasting models. This cooperation leads to accurate and timely data for solar farms, thus further improving the accuracy levels of forecast and operating strategy. The most important application of metadata is to make solar energy systems dynamically configurable. Especially real-time metadata regarding cloud cover and dust storm activity can make solar trackers activate the arrays of solar panels to change angles, thereby reducing losses in efficiency. A typical example from the district of Jodhpur, Rajasthan, shows that such adjustments reduce losses by 10%–15% when wind conditions prevail.

Metadata optimizes battery storage configurations. In peak hours of sunshine, excess energy is stored in the battery, and algorithms based on metadata predict ideal discharge times like peak hours of evening when demand is high. This is highly relevant for decentralized solar grids in rural areas to build reliability.

Metadata is critical in all the Indian solar energy generation steps, ensuring increased reliability and efficiency. In states such as Rajasthan, Gujarat, and Madhya Pradesh, many solar farms rely heavily on metadata to maximize energy productivity. For example, with such significant dependence on the ability of metadata to predict and hence manage dust levels while monitoring temperature fluctuations, Rajasthan Bhadla Solar Park derives much more energy output during elevation periods. Ground-based irradiance sensors, satellite imagery, and analytics driven by artificial intelligence fundamentally support operations, which provide critical insights into both environmental and operational conditions. Data assimilated into machine learning algorithms based on statistics about patterns of sunlight, temperature differences, and wind speed predict energy production and guide real-time decisions. This approach reduces inefficiencies, enhances the effectiveness of the system, and enables predictive maintenance, reducing operation costs and improving grid stability, thereby making metadata an essential component of India's solar power strategies under initiatives like the National Solar Mission.

7.2.2.1 Significance of Metadata

The strategic development of solar farms depends on the analysis of historical and spatial metadata regarding suitable locations. Such metadata would provide considerable insight into pertinent factors such as solar irradiance, topography, temperature anomalies, and the land's suitability, hence permitting data-driven decisions for site selection. Rajasthan offers an interesting case where extensive desert lands with high solar irradiance have increasingly been utilized for the development of large solar parks, like Bhadla Solar Park, that covers over 14,000 acres and produces over 2.5 GW electricity.

7.2.2.2 Metadata and Indian Solar Parks

- **Rewa Ultra Mega Solar Park, Madhya Pradesh**: **I**t is spread across 1590 acres of land and generates 750 MW electricity. Metadata was highly required for site selection as it has a strong solar irradiance and weak cloud cover. This solar park provides electricity to operate the Metro rails in the capital of India, i.e., Delhi.
- **Pavagada Solar Park, Karnataka**: Pavagada spans 13,000 acres and produces 2 GW of power. Metadata land use and weather conditions analysis resulted in this location as it is relatively arid and the agricultural utility of the land is low.
- **Charanka Solar Park, Gujarat**: This is one of the first large solar projects in India, which uses metadata to monitor solar radiation and hence maximizes energy production, producing around 600 MW of electricity.

Metadata-based location selection for solar energy projects relies on critical factors that ensure efficiency and sustainability. Data from satellites and ground-based sensors identify areas with consistent and abundant sunlight, such as Rajasthan's Thar Desert. Land suitability is analyzed through geospatial methods, evaluating gradients, proximity to water sources, and land-use patterns to optimize site selection. Historical climate data, including information on cloudbursts and rainfall, help mitigate risks of power supply disruptions by accounting for environmental challenges. Proximity to existing grid infrastructure is another key factor, as it reduces costs and minimizes energy losses during distribution. By leveraging these data-driven insights, solar parks like those in Rajasthan have significantly enhanced energy production while addressing logistical and environmental hurdles, contributing to the efficient use of land and resources for renewable energy development in India.

The interrelation between metadata and machine learning changes the whole solar system mainly in the direction of efficiency and predictive maintenance. Metadata offers a richly structured dataset to predict patterns and track trends to make informed decisions. The interrelation is critical for solving issues in solar energy, primarily panel degradation, energy losses, and impacts of environment.

7.2.2.3 Metadata and Its Applications

Machine learning models built from historical metadata are highly useful to predict energy losses and determine cleaning schedules. The approaches consider conditions like meteorological conditions, dust deposition rates, and time between cleanings in order to arrive at optimized energy generation with minimum operational costs. For instance, the pilot Rajasthan Bhadla Solar Park has witnessed around 12% efficiency increase for panels after using metadata-driven AI systems that modeled and managed losses from the influence of dust.

Metadata that is containing trends of temperature fluctuations, moisture variation, and actual performances of solar panels is used in machine learning algorithms to forecast the onset of degradation. Such predictive power allows maintenance to be proactive in decreasing downtime and increasing the life cycles of equipment.

Metadata can also tweak angles of panels and adjustments of inverter settings and balance the load distribution within real-time environmental data. Machine learning algorithms enable the detection of anomalies in energy output by comparing real-time data with historical standards, thereby enabling quick resolution of issues.

7.2.2.4 Global and Local Impacts of Metadata

Companies like Google and IBM on the global level have successfully applied artificial intelligence to enhance the efficiencies of solar farms using predictive analytics from metadata. Indian startups such as Prescinto and CleanMax Solar are doing similar things at a local level to monitor the performance of solar and optimize output. This metadata-driven approach is revolutionary in Rajasthan, where challenging environmental conditions require advanced solutions to maintain high levels of energy production.

Besides efficiency in operational work, correct metadata usage in machine learning also has resulted in the generation of more intelligent and responsive solar power systems that would face the changing challenges.

In the state of Rajasthan, where dust collection on solar panels is a major concern, energy loss prediction models developed based on historical metadata employ mechanisms of machine learning to predict losses in energy and recommend cleaning schedules. Such models also consider meteorological conditions, accumulation rate of dust, and frequency of cleaning into account for optimization in energy production with minimal operational cost. For instance, a pilot project at Bhadla Solar Park in Rajasthan registered a 12% gain in panel efficiency when metadata-based AI systems were installed that could simulate and eliminate dust-induced losses. The application of metadata in tapping the potential of solar energy is listed below.

- Degradation signals from metadata on temperature fluctuations, humidity levels etc., help to predict the deterioration of solar panels and take predictive maintenance at the earliest, and this reduces downtime of panels and enhances life cycle.

- AI models trained on process metadata can alter panel orientations, inverter settings, or even in real-time, redistribute loads based on prevailing conditions. Machine learning-based algorithms compare real-time data with historical benchmarks to detect anomalies in energy production that enable quick resolution of problems.
- Machine learning combined with rich metadata makes the operational effectiveness of things smarter and more resilient through adaptable solar energy systems that can get accustomed to changing challenges.

7.2.3 Blockchain in Solar Energy

Blockchain is a digital record that is distributed, decentralized, and systematically noted transactions throughout a network of computers. This block encases every transaction, and the interconnection of these blocks forms a chronological sequence. Once included in the blockchain, the data become irreversible, thus guaranteeing transparency and security [4].

Blockchain ensures that there are free intermediates and low-cost tracking of energy production, trade, and credits in the solar energy. Blockchain-enabled micro-grids will empower P2P energy trading in the distant reaches of Rajasthan. The production and use of energy will be recorded on smart meters and stored on a blockchain network, and surplus energy can even be sold directly to neighbors or anyone else beyond the monopoly of utilities.

Blockchain simplifies carbon credit trading by securely recording renewable energy generation and automating credit allocation. In Rajasthan, solar parks like Bhadla use blockchain to verify energy outputs, ensuring accurate credit issuance too.Blockchain technology [BCT] prevents frauds through immutable records and simultaneously connects the grids around the world.

This integration of metadata and BCT can augment global trust toward Indian renewable energy sector by enhancing transparency and effective management of carbon credits across the nation. Blockchain-based systems have also provided leading support to the implementation of transparent crowdfunding mechanisms for solar projects.

7.3 Integration of IoT, Metadata, and Blockchain

7.3.1 Smart Contracts in Energy Trade

Smart contracts are a great resource for the automation of energy transactions. The term of the agreement is written as code in the contracts to execute a transaction automatically upon occurrence of the predefined conditions. In Rajasthan, the data of real-time energy produced through solar sources are captured through IoT sensors and, for validation of metadata including the intensity of the sunlight, performance of panels, and environmental conditionsare recorded and validated to ensure accuracy of data and efficiency of the system. All the information are held in a blockchain, and the smart contracts enable transaction between energy producers and consumers, thus allowing for quick and transparent exchange.

For example, in Rajasthan's pilot blockchain-based projects in Jaipur, smart contracts have significantly increased transaction efficiency, fully automating payment processing, and energy trading so that transactions are reduced by 50% and intermediaries and administrative overheads are minimized. Blockchain technology ensures the security, transparency, and immutability of all transactions, thereby substantially reducing fraud and disputes. This optimized methodology reduces the operational expenditures of solar energy producers while increasing the effectiveness of P2P energy trading between households and minor producers as well, to sell their surplus solar energy directly to other members of the community.

IoT, metadata, and blockchain confluence to create an efficient solution to the problem of power and planet. Example: Rajasthan's solar parks, namely Bhadla Solar Park: The real-time data from IoT sensors, are fed into AI algorithms that will predict energy demand and supply fluctuations and optimize grid stability along with resource allocation.

Huge volume of data generated by the sensor related to temperature, solar irradiance, and performance of panels, artificial intelligence can predict such times when energy demand exceeds an average value. For instance, AI improves inverter functions, solar tracking systems, and battery storage units for real-time optimization, and these will distribute energy more effectively during peak times. The preemption will therefore be beneficial for the solar energy storage solutions because it minimizes the resultant waste of the surplus energy generated during intervals of higher solar irradiation.

AI-optimized algorithms can be used in preventive maintenance by detecting early signs of operational inefficiencies or potential failures and thereby decreasing downtime and maintenance costs.

This will further assist the AI system to adjust more accurately the solar system configuration in accordance with environmental conditions such as dust deposit or seasonal weather conditions, which may cause a wide variation in the performance of solar panels in deserts like Rajasthan.

7.3.2 A Case Study

Bhadla Solar Park in Rajasthan is the world's largest solar facility. It's a great example of how IoT, metadata, and blockchain technologies can be combined to make the production of solar energy more productive. The solar park covers more than 14,000 acres and produces more than 2.5 GW of electricity; the fact is that this has significantly added to the renewable energy capacity of India.

Bhadla through IoT devices ensures that solar panels can be continuously monitored for their performance, monitoring real-time energy output, temperature, and degradation of the panels. The data collected by these devices are connected to a central database that analyzes the metadata in terms of optimizing energy production and predicting future performance trends.

Blockchain technology happens to be an application that enhances the transparency and security of energy transactions from generation all the way down to the point of sale to consumers or the electrical grid. With all the energy transactions recorded in a non-alterable ledger, it avails to all the stakeholders, i.e., energy producers, consumers, or regulatory bodies. Bhadla project has successfully reduced the operational costs associated with panels to facilitate the future growth of this park together with more-efficient energy management and easier detectability.

7.4 Conclusion

The integration of IoT, metadata, and blockchain in the state of Rajasthan is creating an effective ecosystem for efficient energy generation, management, and trade in the solar energy sector. Such technologies shall optimize the solar energy systems with real-time monitoring, predictive maintenance, and secure decentralized transactions, to stabilize the grid and slash costs. For instance, case studies at Rajasthan including Jaipur and Bhadla reflect the promise of blockchain-enabled smart contracts in terms of energy sales efficiency with up to 50 percent reduction in transaction times, whereas AI-driven optimization improves energy efficiency and reliability of the system. This is not only the progress seen in Rajasthan but has the potential to be scaled up all over India and beyond to help the country reach its aspirational target of 500 GW of renewable energy by 2030. Also, blockchain technology that India is incorporating into its "One Sun, One World, One Grid" initiative can help set up an interconnected global energy network where energy transactions take place seamlessly across national borders, thereby reducing reliance on fossil fuels and accelerating the world's transition toward a cleaner and greener energy future.

References

1. Sarath, K., Guttikunda, Puja Jawahar: Atmospheric emissions and pollution from the coal-fired thermal power plants in India. Atmos. Environ. **92**, 449–460 (2014), ISSN 1352-2310. https://doi.org/10.1016/j.atmosenv.2014.04.057
2. Mondal, T., Sengupta, D., Mandal, A.: Natural radioactivity of ash and coal in major thermal power plants of West Bengal, India. Curr. Sci. **91**(10), 1387–1393 (2006). http://www.jstor.org/stable/24094063
3. Vig, N., Ravindra, K., Mor, S.: Environmental impacts of Indian coal thermal power plants and associated human health risk to the nearby residential communities: a potential review.

Chemosphere **341**, 140103, ISSN 0045-6535 (2023). https://doi.org/10.1016/j.chemosphere.2023.140103

4. Vionis, P., Kotsilieris, T.: The potential of blockchain technology and smart contracts in the energy sector: a review. Appl. Sci. **14**(1), 253 (2023). https://doi.org/10.3390/app14010253
5. Global Energy Policy 2013, The policy institution for Global Energy thought Leadership. https://www.energypolicy.columbia.edu/about-us/
6. Kumari, O., et al.: Environmetal kuznet's curve for sustainable development. Int. J. Innov. Technol. Exploring Eng. **12**(s), 813 (2019)
7. Annual Report 2018, Central Electricity Authority. http://www.cea.nic.in/reports/committee/nep/nep_dec.pdf
8. Oliveira et al.: Environmental application of biochar: current status and perspectives. Bioresource Technol. **246**, 110–122, ISSN 0960-8524 (2017). https://doi.org/10.1016/j.biortech.2017.08.122
9. Oum, K.R.: Empowering energy transition through block chain technology. In: Rehman, A., Selvam, M., Acharya, B.R. (eds.) Blockchain and Cryptocurrency: Management Systems and Technology Challenges. CRC, UP (2024)
10. Zhou et al.: Total concentrations and sources of heavy metal pollution in global river and lake water bodies from 1972 to 2017. Global Ecol. Conservation 22, e00925, ISSN 2351-9894 (2020). https://doi.org/10.1016/j.gecco.2020.e00925
11. Miglani, A., Kumar, N., Chamola, V., Zeadally, S.: Blockchain for internet of energy management: review, solutions, and challenges. Comput. Commun. **151**, 395–418 (2020)

Chapter 8
Smart Tourism Ecosystems: Integrating IoT, Blockchain, and Metadata for a Connected Future

Devvandya Mishra and Lokesh Vijayvargy

Abstract The tourism industry is undergoing an exceptionally rapid transformation driven by technological advancements. It continues to bring the world closer and more accessible, making travel and adventure activities more interesting and accessible for all, and also ensuring efficiency and interconnectedness. This book chapter explores the concept of the smart tourism ecosystem and the role of advanced technologies such as the Internet of Things, blockchain, and metadata in creating a more effective, more connected, personalized tourism experience, and a secure ecosystem. It will examine how this advanced technology is reshaping the tourism industry in various aspects, such as destination management and effective service delivery to tourists, and ensure a high level of user satisfaction. This chapter will also discuss the challenges and limitations of smart tourism, which include technological dependencies, data security and privacy concerns, and interoperability issues, and will highlight the need for a proper, robust policy and governance framework. By analyzing real-world applications and future trends of these technologies, this chapter's objective is to provide valuable and in-depth insight for academicians, policymakers, researchers, and other stakeholders in the tourism industry.

Keywords Smart tourism · Internet of Things · Digital transformation · Metadata · Blockchain technology

8.1 Introduction

Tourism is one of the fastest-growing industries in the world, and it plays a significant role in spreading globalization, cultural exchange, and economic development. Tourism is an artistic, financial, and social activity that engages people in travelling to

D. Mishra (✉) · L. Vijayvargy
Jaipuria Institute of Management, Jaipur, Rajasthan, India
e-mail: devvandya.mishra.fpm23j@jaipuria.ac.in

L. Vijayvargy
e-mail: lokesh.vijayvargy@jaipuria.ac.in

P. M. Jeyanthi (ed.), *Distributed Ledger Technology in Communication*, Signals and Communication Technology, https://doi.org/10.1007/978-3-032-07682-3_8

destinations outside their everyday surroundings for their enjoyment or professional reasons [50]. Tourism encompasses people travelling from one place to another, staying outside their hometown or usual environment for various reasons such as business, pleasure, exposure, adventure, leisure, etc. It involves the movement of people and all the other industries and services that support and promote these activities, such as accommodation, attractions, transportation, food and beverages, and other hospitality activities. The tourism industry is committed to creating and facilitating a unique, memorable, and rich experience for travelers [24]. It is also true that the conventional method of tourism is insufficient to attract enough tourists to revitalize the economy efficiently [33]. It naturally faces various types of challenges, such as political, economic, environmental, and other geopolitical issues, as well as pandemics [35]. Technology has emerged as a boon after facing many challenges, and the widespread impact of the pandemic has affected every sector [33]. With technological advancements reshaping every industry, the tourism sector has also advanced technologically, which is now known as smart tourism. Smart tourism is seen as a revolution in the tourism industry, involving transformative and innovative theoretical and practical approaches for this sector [33].

Smart tourism is not just about the use of technology; it's also involved in transforming the traditional tourism system into an intelligent and tech-savvy ecosystem, where all the data and information flow freely among travelers, service providers, and various government agencies to optimize resource allocation, visitor satisfaction, and decision-making. In other words, it refers to applying advanced information technologies to enhance resource management, the tourism experience, ensure sustainability, and maximize destination competition. Smart tourism is a significant shift in the tourism industry, which drives the integration of all advanced technologies to create personalized, customized, and enhanced tourist experiences [22]. Advanced technologies include integrating upgraded technologies like the Internet of Things, blockchain, big data, artificial intelligence, and efficient travel experiences. The incorporation is not merely about adopting new equipment, gadgets, applications, and software, but about creating a holistic ecosystem where technology seamlessly enhances every aspect of the tourist journey, from planning, booking, experience, and sharing. It transforms traditional tourism services into interconnected, data-driven ecosystems that enhance convenience, satisfaction, and sustainability for service providers and travelers alike.

At the heart of smart tourism, there are several core technological components, which include big data analytics, the Internet of Things, and artificial intelligence [22]. Beyond just technology, smart tourism encompasses a broader range of factors, including technological, social, and economic aspects, which impact how the destination is managed, how business operates, and how tourists interact with their environments. This could be at the micro and macro levels [54].

8.2 Origin of Smart Tourism

The concept of smart tourism evolved in the early 2000s. It is founded on two key references, the first one being the concept of smart cities, and the other the evolution of e-tourism [47].

The evaluation of smart tourism involves the progressive integration of various cutting-edge technologies, which makes travel easier, more connected, and smarter for everyone, from planning a trip to experiencing a destination [39]. In the initial stage, smart tourism starts with simple online booking on a website and travel information. Then, it shifted to mobile applications and social media to help travelers share and plan more easily. After that, it used advanced technologies like artificial intelligence, the metaverse, AR, VR, XR, and blockchain technology to make travel more personalized, fun, and innovative [17, 39]. All these advanced technologies work together to provide a complete and engaging travel experience. For example, IoT equipment can collect real-time data on travelers' moments and help measure environmental conditions. Artificial intelligence can be beneficial in personalized recommendations and services based on individual preferences. Besides, blockchain technology is the best data storage option; big data analytics can identify new trends and patterns to improve destination management. The combination of all these technologies truly shapes the innovative tourism ecosystem. The smart tourism agenda not only enhances tourist experiences but also increases profitability for the business and generates employment on a mass level, ensuring that tourism practices are environmentally responsible and contribute positively to the local communities. Sustainability is the key element of innovative tourism activities, promoting cultural preservation and supporting local economies. By adopting this holistic approach that considers the needs of travelers, businesses, and local communities, smart tourism strives to create a more resilient and equitable tourism industry that can thrive in the long term.

8.3 Significance of Technology in Reshaping the Tourism Industry

The tourism industry is undergoing a rapid transformation as technological advancements continue to bring the world closer, making travel and adventure more accessible, interconnected, and efficient [2]. Technology plays a very significant role in reshaping the tourism industry, streamlining the operations and users' personalized experience, and providing valuable insights for adequate destination management, from simplifying the booking and payment process through various online platforms and mobile applications to offering real-time updates and information regarding weather, destination, technology, and transportation, which ensures efficiency and convenience for both service providers and tourists [1]. It acts as a catalyst for innovation and enables tourism-related businesses to provide more efficient, creative, engaging services to the customer. Today's tourists are exploring new destinations and looking for a smooth and hassle-free experience. The availability of sophisticated infrastructure, equipment, and state-of-the-art technological amenities influences these tourist experiences [42].

The blend of traditional with modern, the historical with futuristic, sets a successful tourist destination apart. When historical sites adopt digital storytelling, when cities implement digital and smart transportation or nature trails with digital guides, it will create a seamless mix of modernity and authenticity that will enrich travel experiences and bring a new image of travel [3].

8.4 Role of IoT, Blockchain, and Metadata in Tourism Digitalization

Technologies like blockchain, artificial intelligence, metadata, and the Internet of Things drive the digital transformation in the tourism sector. The Internet of Things is somehow a leading path to a smart world with ubiquitous computing and networking. It aims to simplify different tasks and provide another task, such as monitoring the different phenomena surrounding us. IoT-based tourism services will automate various tourism tasks, such as booking, allowing for bright hotel rooms, automated check-in, real-time traffic updates, personalized service, and connected objects to build a smarter world and smart tourism. This is where a specific application will monitor the environment and automatically respond, for example, minimize energy wastage, ensure appropriate planning, and be helpful in crowd management [9]. The integration of IoT will enhance convenience, efficiency, and safety for both service providers and tourists. Blockchain is a distributed, decentralized, and public digital ledger. Using blockchain technology in the tourism industry will ensure high-level security, transparency, and transaction trust by enabling secure payment methods, verifiable reviews, and digital identity management [40]. Meanwhile, metadata plays a very significant role in categorizing and organizing a vast amount of travel-related information and data. By enabling more accurate search results, content delivery, and recommendations, metadata ensures that travelers receive timely, reliable, relevant, and personalized information. Together, all these technologies are redefining how tourism is managed, experienced, and marketed in the digital age.

The primary objective of this chapter is to examine how integrating innovative technologies, specifically metadata, blockchain, and the Internet of Things, transforms the tourism industry into a more secure, connected, modern, personalized ecosystem. By examining and investigating these advanced technologies' role and collective synergy, the basic objective is to highlight their potential for enhancing the tourist experience, promoting sustainable and innovative tourism practice, and optimizing operations. This chapter also seeks to provide practical insights through real-world applications, while addressing the current challenges and offering future directions to academicians, policymakers, researchers, and other stakeholders.

8.5 Understanding the Technologies

8.5.1 Internet of Things (IoT) in Tourism

The Internet of Things (IoT) was initially introduced by the Auto-ID research Centre at the Massachusetts Institute of Technology (MIT), where a significant effort was made to identify products uniquely [9]. The Internet of Things refers to physical objects equipped with sensors that communicate with computers and other electronic devices, such as mobile phones, tablets, etc., and enable the physical world to be digitally controlled, monitored, and observed [31]. It uses various types of technology to connect the digital world to the physical world, such as sensors, actuators, etc. The constant connectivity enabled by Internet of Things (IoT), along with various other equipment such as the power of data and analytics, opens new opportunities and avenues for the industry or companies to innovate and transform their services, products, and systems, also enhancing their operational efficiencies and effectiveness [31]. In the current scenario, the trend of IoT can be seen in every industry, whether in manufacturing or services. The potential of IoT is vast, and it is continuously expanding. By 2030, its global value is estimated to reach as much as $12.5 trillion, which encompasses the profit gained by industries, service providers, customers, and final users. The IoT has transformed every industry, and the tourism industry is also enhancing operational efficiencies and visitor experience. The application of IoT could be seen in the tourism industry, such as smart hotels and rooms that are connected with devices for automated room settings, applying an intelligent transportation system that optimizes travel routes, tourism tricks for security and safety, and resources, crowd management, and providing personalized services tailored to individual preferences and personalized services. Apart from that, real-world applications of IoT can be seen in airports, where it supports efficient baggage handling and updates real-time flight routes, and museums, where IoT enables guides and interactive displays to increase the visitor experience. This application improves tourist convenience, enhances satisfaction, helps service providers provide data-driven services, and makes them more responsive.

Figure 8.1 highlights the six critical parameters that form the foundation of the modern digital and smart tourism ecosystem. The enterprise and consumers must meet the following six requirements to achieve a seamless, smooth, and effective IoT [31].

8.5.2 Blockchain in Tourism

Blockchain refers to a decentralized and unchangeable data storage system that ensures the collection, recording, and tracking of information and data within a shared and transparent ecosystem. It is a distributed ledger technology (DLT) that enables the secure, transparent, and tamperproof storage of information across a

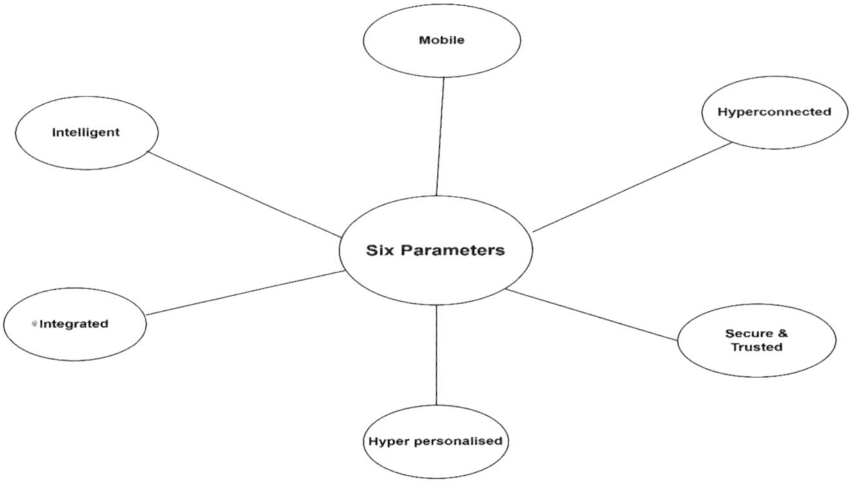

Fig. 8.1 Six key parameters of a digital ecosystem

network of connected computers, known as nodes. This technology creates a digital record of transactions or information, data that is grouped into linked blocks sequentially in chronological order, called a blockchain [49]. In a blockchain, each block contains a set of data, such as transactions, records, and other types of information. Blockchains act as trust-building systems that promote transparency, appropriate support, and facilitate conditions for smooth and seamless information exchange [53]. The tourism industry plays a significant role in the global economy. Before the pandemic in 2019, travel and tourism contributed 10.4% to the worldwide GDP, equivalent to US$ 9.2 trillion, and supported 10.6% of total employment worldwide [52]. Blockchain technology has the potential to revolutionize the tourism industry [46]. Blockchain technology empowers tourism firms, specifically smaller ones, to access consumer data directly from users, instead of relying on and purchasing directly from social media sites such as Instagram, Facebook, Google, etc. With the help of this technology, tourism sector firms can purchase data ethically, directly, and transparently from consumers, which gives them more control over their data [44]. Blockchain technology also makes the tourism sector more intelligent, for example, with the combination of big data analytics and blockchain, artificial intelligence could enhance the understanding of tourist needs and wants, enabling customized services for tourism products and services [23]. Blockchain can automate, streamline, and secure check-in processes at airports and hotels, enabling the use of smart gates, face scanners, digital passports, digital keys, and other biometric identification for seamless and automated entry [45]. Blockchain also converts the physical ecosystem to a digital ecosystem and can be managed virtually, and it is changing the way people live, work, travel, and do business, and in the process, transforming and reshaping the entire tourism industry [8]. Blockchain offers a unified platform for tourists to connect and coordinate travel and tourism activities, which improves user experience and privacy, security, where all price information and other aspects are continuously updated [32]. The blockchain technology has also reduced the need for or the level of intermediaries in the tourism sector. Eliminating intermediaries increases transparency, accountability, collaboration, coordination, and trust among all stakeholders in the tourism sector [37].

8.5.3 Metadata in Tourism

Metadata refers to information such as author, creation date, and file size describing a data point or set. It also improves the data system's functions and operations and makes it easier to search, organize, manage, and use the data [7]. In simple terms, it is data about data, which provides details about the data that is separate from the content of the data itself [7]. Metadata is also crucial in data management and governance, as it is helpful for data equality and integrity. In the tourism context, metadata refers to providing information about other data, helping collect, organize, manage, and retrieve travel-related content more efficiently and effectively. It plays a crucial role in enhancing the digital experience by enabling personalized recommendations, more accurate search results, and seamless integration with booking platforms, travel guides, and location-based services. By improving data usability and accessibility, the metadata supports a smarter decision-making process and a more connected and innovative tourism ecosystem.

Figure 8.2 showcases the integration of Internet of Things, blockchain, and metadata in the tourism industry, creating a powerful synergy that will amplify the strength of each technology. The real-time data generated by the Internet of Things, with the help of tourist interactions, their preference, location, and metadata structure, is categorized for personalized services. Blockchain technology will ensure data transparency, authenticity, and security, enabling trusted transactions and safeguarding consumers' personal information [48]. These technologies will enable a seamless experience, such as real-time updates, a personalized travel experience, and recommendations. This synergy will make the tourism sector faster and smarter and connect it more innovatively, benefiting customers and service providers.

8.6 Proposed Model for Smart Tourism Ecosystem

Figure 8.3 highlights the interconnected roles of tourists, service providers, governance authorities, and the technology layer in the smart tourism ecosystem. The technological layers act as the central enabler, facilitating real-time data exchange, seamless communication, and efficient coordination among stakeholders to enhance tourist satisfaction, experience, and operation.

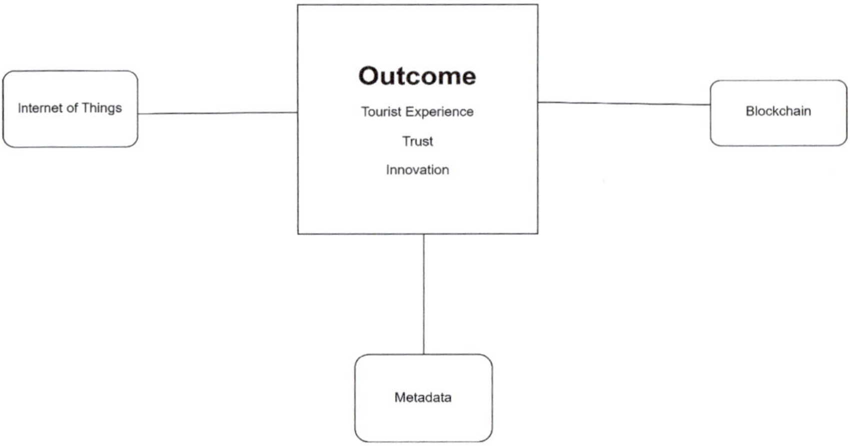

Fig. 8.2 Integrated technological framework: leveraging IoT, blockchain, and metadata in tourism

Fig. 8.3 Proposed smart tourism ecosystem model

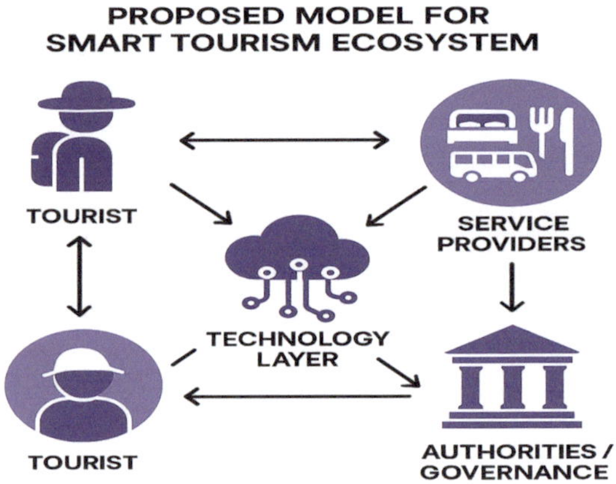

8.7 Application in the Tourism Sector

8.7.1 Smart Destination Management

Smart destination management refers to the intelligent core of a thriving, innovative tourism ecosystem, leveraging real-time data collection, recording, and predictive analytics [20]. The constant flow of data from different sources, such as social media, mobile apps, sensors, and other transport systems, provides insight into what's happening at the destination. This allows destination management organizations (DMOs) and stakeholders to gain a comprehensive understanding of tourist behavior, environmental impact, and infrastructure performance. For instance, real-time data from sensors placed throughout a city can monitor pedestrian traffic, public transportation usage, and parking availability, and enable dynamic adjustments to traffic flow and resource allocation [19]. Mobile applications can provide personalized recommendations to tourists based on their preferences and location, offering information on nearby attractions, events, and services while collecting valuable data on their interests and investments. Furthermore, innovative destination management plays a very significant role in enhancing the visitor experience. By leveraging data analytics, DMOs can identify trends and personalized offerings, ensuring that tourists receive relevant and appropriate information [41]. An augmented reality application could overlay historical information onto real-world landmarks, enriching the sightseeing experience. The intelligent tracking system can streamline access to attractions and transportation, reducing wait times and improving overall satisfaction.

Sustainability is the central key pillar of innovative destination management. By inspecting waste management, monitoring energy consumption, and monitoring environmental conditions through connected devices, destinations can make very informed decisions to minimize their ecological footprint [4]. A smart grid can optimize the distribution of energy, while an intelligent waste management system also improves collection and ensures recycling and efficiency [5]. Data on visitor movement patterns can also inform strategies to manage the crowd and protect sensitive natural areas. Some real-time examples of innovative destination management: Amsterdam utilizes intelligent sensors to monitor the canal water quality and optimize traffic, while Barcelona employs a city-wide network of sensors to manage the waste, ensuring the quality of residents and visitors. In India, initiatives like innovative city projects are beginning to incorporate elements of smart tourism, focusing on digital infrastructure, e-governance, and citizen services that can benefit local people and tourists.

8.7.2 Enhanced Customer Experience

In the evolving landscape of destination management, the primary focus is increasingly shifting from sampling to attracting visitors and cultivating truly exceptional and memorable experiences. Today, technology empowers us to craft journeys as unique and adventurous as those who embark on them. Customized tours represent a departure from the standardized approach, which allows visitors to actively participate in shaping their exploration of our rich culture and heritage. Also, AI-based recommendations are intelligent systems that analyze the vast dataset to anticipate visitor preferences and offer timely and relevant suggestions. The synergy between customized tours and AI-driven insight creates a compelling ecosystem for tourist to enhance their experience and satisfaction [25]. The artificial intelligence identifies potential interests that visitors might not even be consciously aware of, leading to the creation of truly bespoke tour options. That's why continuous feedback from the customized tours enhances the AI's understanding of tourist preferences and likes and dislikes, making future recommendations even more precise, relevant, and systematic [43].

8.7.3 Safety and Security Enhancement

In the era of increasing global interconnectedness, ensuring the safety and security of tourists is one of the main objectives for service providers. Innovative destination management provides cutting-edge technologies to create a secure and trustworthy environment, enhancing the overall travel experience and fostering tourist confidence. At this time, various service providers are using blockchain technology to store travelers' essential information, such as verified identification, medical records, tourist insurance details, etc. This decentralized and temperature-proof system provides various advantages. For tourists, it means a single, secure digital identity that can streamline all airport procedures, restaurant and hotel check-in, and even access to certain services, all while maintaining control over the personal data. In the case of an emergency, authorized personnel can quickly and securely access all crucial information, which facilitates faster and more effective assistance. For the destination management, these advanced technology offers enhanced capabilities to reduce the risk and improve data integrity, contributing to a safer overall environment. Apart from that, wearable devices or smartphone applications equipped with the emergency SOS features can allow visitors to instantly request assistance, transmitting their precise location to emergency responders.

8.7.4 Sustainable Tourism

Sustainable tourism is increasingly assisting advanced technologies to reduce the negative environmental impact, and apart from that, it is also enhancing accountability and transparency [51]. Smart destination management offers an innovative solution to minimize our ecological footprints, also promotes responsible travel with energy monitoring via Internet of Things and transparent carbon credits via blockchain, standing out as a powerful tool in the endeavor [29]. These advanced technology helps in understanding the energy consumption level or patterns across the destination infrastructures from hotel transportation networks to iconic landmarks. These all-connected devices provide real-time data for the electricity usage, enabling informed decisions about optimizing energy efficiency, identifying areas of wastage, and promoting the adoption of renewable energy sources [14]. By using smart sensors and connected devices, all tourist operators can easily identify the variances and inefficiencies, optimize energy use, and significantly contribute to carbon emission reduction. Data-driven approaches not only reduce the operational cost but also drastically reduce the level of carbon emissions and carbon footprint, which is associated with tourist activities, contributing to a greener and more sustainable tourist destination [28]. These practices not only build trust among tourists related to environmentally conscious decisions but also encourage tourism service providers to adopt green practices in tourism activities. Together, these advanced technologies contribute to a more responsible and sustainable tourism industry.

8.8 Challenges and Limitations

8.8.1 Dependency on a Technological Pyramid

The technological pyramid refers to a layered and interconnected digital infrastructure required for a seamless, effective, and intelligent travel experience [34]. At the base of this pyramid, the hardware is at the bottom, including sensors, servers, smartphones, and other smart devices, which form the physical foundation of the innovative tourism system. After that, the next layer includes software applications that process the data, provide a user interface, and also enable online bookings, personalized recommendations, and real-time navigation. Above this layer, there is high-speed internet, 6G networks, and cloud computing, which actually ensures uninterrupted access and data transmission between devices, platforms, and users [27]. At the top of the pyramid, the smart tourism experience itself is found, where all the technologies come together to offer service to tourists based on their choices, and it will be entirely based on customization, engagement, and convenience. This pyramid shows that smart tourism cannot function in isolation; all layers are required for seamless work. A breakdown and discontinuity at any layer, such as a lack of device availability, poor internet connection, or outdated software, can compromise the overall experience of the tourist.

8.8.2 Data Privacy and Cybersecurity Concerns

Smart tourism offered a enhance travel experience and also improved destination management through various advanced technology such as Internet of Things, artificial intelligence, metaverse, blockchain technology, etc., which raises significant cybersecurity and data security, data privacy related concerns due to the extensive collection of tourist personal data, geolocation, behavioral preference, their travel behavior biometric data, financial details and including demographic information [18]. The lack of transparency regarding data usage and potential for unauthorized access or third-party access and misuse, profiling risk, and complexities of varying global data protection regulations erode trust. Simultaneously, the interconnected innovative tourism system expands the attack surface, making it vulnerable to data breaches, ransomware attacks, phishing, malware, DDoS attacks, and threats targeting IoT devices and the entire operations system, supply chain. The potential consequences of cyber-attacks could be financial losses, image damage, loss of goodwill, service disruption, and safety risks, necessitating robust data protection measures, adopting strong cybersecurity protocols, and focusing on effectiveness and transparency to ensure trustworthiness, sustainability, and smart tourism [15].

8.8.3 Interoperability

In a smart tourism system, various technologies or platforms are used, such as IoT, AI, metaverse, blockchain, and machine learning, to provide tourism services seamlessly to tourists, but interoperability challenges pose a significant challenge to service providers [36]. Smart tourism basically relies on the integration of various components such as accommodation services, transportation networks, Internet of Things devices, destination management, and other mobile applications or software. However, all these components were developed by different vendors, that is why there is a lack of standardized protocols, data formats, integration information, technical standards, and communication protocols, leading to fragmentation and incompatibility in the system. As a result, users may encounter difficulties and challenges when they try to access services, such as using single travel applications and websites for booking accommodation, food and beverages, transportation, and checking into hotels, due to a lack of a uniform interface between systems. The lack of interoperability not only affects the tourist experience but also affects the data sharing and real-time decision-making among various stakeholders such as tourism service providers, policy makers, broadband operators, local authorities, and local communities as well. Furthermore, this inconsistent system integration could be increasing the maintenance cost to service providers, reducing scalability, particularly for small and medium-sized firms or enterprises that may struggle to adopt compatible technologies.

8.8.4 Lack of Awareness

In the smart tourism system, there is another significant challenge: a lack of awareness about the availability of smart technologies and their features. Many tourists, local communities, and small service providers are not aware of the advanced technologies now used in the tourism sector, and they are also not aware of the features, uses, benefits, and other functions. This lack of awareness among the users leads to underutilization of these smart technologies, such as travel applications, software, smart guides, real-time access to data, and online booking systems. As a result, many users and tourists may miss out on accessing the personalized service, experience, and other small service providers who are not aware that they fail to optimize their operations and are also unable to reach a broader audience [11]. Moreover, without proper awareness about these smart technologies, investors are also not interested or reluctant to invest in them because they view them as unnecessary or too complex.

8.8.5 Regulatory and Policy Gaps

The technology in the tourism sector is increasing very rapidly, and many existing legal and regulatory frameworks need to keep pace. This creates uncertainty and hurdles in the data handling, resolving conflicts, privacy rights of the users, cybersecurity standards, and ethical uses of emerging technologies such as artificial intelligence and facial recognition. Apart from that, in many countries, there is no clear policy guidance for integrating smart technologies into environmentally sensitive areas, leading to conflicts between conservation efforts and technology development [6].

8.8.6 Environmental and Ethical Concerns

The smart technology's objective is to enhance the tourist experience and satisfaction level, but its widespread use raises various environmental issues. The continuous proliferation of sensors, the Internet of Things, and other digital infrastructure in tourist destinations can contribute to high energy consumption and also electricity waste. Apart from that, the disposal of outdated electronic equipment and devices waste from the tourism industry can lead to environmental pollution if they are not properly recycled, reused, and contribute to the growing problem of electronic waste [26]. Ethically, the use of surveillance technologies such as real-time tracking systems and facial recognition raises concern [30]. Tourists and service users may not be fully aware of the extent to which their personal information is being monitored and stored, leading to debates over surveillance versus individual freedom. Moreover, the digital divide can create exclusion, as specific populations, especially in developing countries in rural areas, may not have access to the internet and technologies needed to benefit from smart tourism solutions.

8.9 Future Directions and Best Way Forward

The future of smart tourism is poised for transformative growth, which is driven by cutting-edge technologies and evolving traveler expectations. In the tourism industry, advanced technologies have significantly shaped the entire industry, such as artificial intelligence, metaverse, Internet of Things, and blockchain technology. The AI algorithm will process the vast amount of data generated by IoT sensors and various devices deployed across numerous destinations, enabling more sophisticated personalized recommendations and predictive analytics for resource management, such as crowd control, energy optimization, and enhanced operational efficiency in the tourism industry. Blockchain technology will address various crucial aspects, such as a transparent transaction system, a secure database, identity verification, fostering greater trust, and security within the entire tourism industry or ecosystem.

8.9.1 Rise of Metaverse in Tourism

In recent years, the travel industry has seen a massive digital shift and technological advancement. Metaverse is seen as the next big revolution in the tourism industry, and it is a collective space where the digital and physical worlds converge and deliver more immersive, intense virtual reality and augmented reality experiences to tourists [12]. Metaverse is a virtual 3D space where users can communicate with other users as well as create avatars and objects [38]. The emergence of the metaverse in the tourism sector will bring new opportunities for the tourism industry and will reshape the entire industry. The metaverse tourism industry includes immersive virtual experiences of destinations, attractions, attention, and even historical events [10]. This could allow potential tourists and users to "try before they buy," experiencing the destination virtually before making a physical trip or physical check [38]. This technology also opens the door for virtual tourism experiences for those individuals who are not able to travel physically and explore the place in real time. Most importantly, the upcoming generation will not be able to see the Egyptian Pyramids in their entirety. However, with the help of the metaverse, it is possible to replicate the reconstruction of all historic structures to their original formal beauty [38].

8.9.2 Policy and Governance Needs

As smart tourism evolves very drastically, there is a pressing need for robust policy and governance frameworks to address the associated challenges. These include clear guidelines for data protection, ensuring ethical use of artificial intelligence, and promoting interoperability among various systems [13]. Governments and all regulatory bodies and agencies will need to adapt existing laws and regulations and required modifications as per the requirement [21]. Also, create some new laws to govern, collect, and use tourist information data and its storage, as well as to ensure fair competition and customer protection in the digital tourism landscape, thereby promoting transparency, data security, and ethical innovation within the entire tourism ecosystem [2, 13].

8.10 Conclusion

The integration of smart and advanced technology in the tourism sector is not just a trend but also a transformative force that is reshaping how the destination operates and how travelers engage with it. From the Internet of Things to AI-powered personalization systems and smart infrastructure to seamless digital services, this smart technology is redefining the tourism experience at every level. This integration of technology is building the tourism sector in this way, so that in the future tourism will be considered as prosperous, responsible, and sustainable. By connecting the data system, aligning the various environmental efforts, and fostering genuine partnership, the tourism industry unlocks efficiencies and effectiveness, enhances tourist experience, and also minimizes the negative impacts. Ultimately, the path of achieving the sustainable, efficient, and truly visitor-centered tourism models is paved with technologies but driven by the vision of interconnectedness [16]. Embracing these integrated solutions from smart destination management systems to circular economy principles will empower the entire tourism industry, benefiting all stakeholders and tourists.

References

1. Abdelfattah, F., Al-Alawi, A., Abdullahi, M.S., Salah, M.: Embracing the industrial revolution: the impact of technological advancements and government policies on tourism development in Oman. J. Policy Res. Tourism Leisure Events (2023)
2. Abdelfattah, F., Al-Alawi, A., Mohammed, M.S.A.: Embracing the industrial revolution: the impact of technological advancements and government policies on tourism development in Oman. J. Policy Res. Tourism Leisure Events (2023)
3. Abdelfattah, F., Halbusi, H.A., AL-Brwani, R.M.: Influence of self-perceived creativity and social media use in predicting E-entrepreneurial intention. Int. J. Innov. Stud. **6**(3) (2022)
4. Alami, T., Aria, T.T.: Sustainable and Smart Destination Management. Umeå School of Business and Economics (2016)
5. Alsharif, A., Isa, S.M., Alqudah, M.N.: Smart tourism, hospitality, and destination: a systematic review and future directions. J. Tourism Services **15** (2024)
6. Amore, A., Hall, C.M.: Tourism Interventions. Routledge (2024)
7. Badman, A., Kosinski, M.: IBM.COM (2025) [Online] Available at: https://www.ibm.com/think/topics/metadata
8. Balasubramanian, S., Sethi, J.S., Ajayan, S., Paris, C.M.: An enabling Framework for Blockchain in Tourism. Information Technology & Tourism (2022)
9. Chaouchi, H.: The Internet of Things. Wiley (2017)
10. Chen, Z.: Beyond boundaries: exploring the metaverse in tourism. Int. J. Contemp. Hospitality Manage. (2025)
11. Cimbaljević, M., et al.: Employees' technology adoption in the context of smart tourism development the role of technological acceptance and technological readiness. Eur. J. Innov. Manage. (2024)
12. Constantin, M., Genovese, G., Munawar, K., Stone, R.: Mckinsey.com (2023). [Online] Available at: https://www.mckinsey.com/industries/travel/our-insights/tourism-in-the-metaverse-can-travel-go-virtual
13. Edgell, D.L., Swanson, S.J.R., Allen, M.D., Smith, G.: Tourism Policy and Planning (2008)

14. Fatema, K., Sinnappan, P., Meng, C.S., Watabe, M. (2024). Technological advancements and innovations in the tourism industry driving sustainable tourism. In: The Need for Sustainable Tourism in an Era of Global Climate Change: Pathway to a Greener Future (2024)
15. Florido-Benítez, L.: The role of cybersecurity as a preventive measure in digital tourism and travel: a systematic literature review. In: Discover Computing (2025)
16. Forsyth, T.: Environmental responsibility and business regulation: the case of sustainable tourism. Geographical J. (1997)
17. Garg, A., et al.: Examining the smart tourism landscape: a bibliometric study and analytical synopsis of IoT research. In: 2024 International Conference on Intelligent & Innovative Practices in Engineering & Management (IIPEM) (2024)
18. Ghaderi, Z., Beal, L., Houanti, L.: Cybersecurity threats in tourism and hospitality perspectives from tourists engaging with sharing economy services. In: Current Issues in Tourism (2024)
19. Gretzel, U.: The smart DMO A new step in the digital transformation of destination management organizations. Eur. J. Tourism Res. **30** (2021)
20. Ivars-Baidal, J.A., Celdrán-Bernabeu, M.A., Mazón, J.-N., Perles-Ivars, Á.F. (2019). Smart destinations and the evolution of ICTs: a new scenario for destination management? Current Issues Tourism **22**(13) (2019)
21. Khater, M., Ibrahim, O., Sayed, M.N.E., Faik, M.: Legal frameworks for sustainable tourism: balancing environmental conservation and economic development. In: Current Issues in Tourism (2024)
22. Krishnamma, B., Babu, K., Zacharia, A.M.: Tourism 2.0: exploring the evolution of smart destinations. J. Manage. Res. Anal. **12**(1), 3–9 (2025)
23. Leal, F., Malheiro, B., Veloso, B., Burguillo, J.C.: Responsible processing of crowdsourced tourism data. J. Sustain. Tourism (2021)
24. Leung, X.Y.: Technology-enabled service evolution in tourism: a perspective article. Tourism Rev. (2020)
25. Liberato, P., Alen, E., Liberato, D.: Smart tourism destination triggers consumer experience: the case of Porto. Eur. J. Manage. Bus. Econ. **27**(1) (2018)
26. Li, C., Agyeiwaah, E., Zhao, Y.: Understanding tourists' eco-paralysis, environmental concern, and pro-environmental behavior: an explanatory sequential mixed method study. J. Sustain. Tourism (2023)
27. Li, Q., et al.: 6G Cloud-Native System: Vision, Challenges, Architecture Framework and Enabling Technologies. IEEE Access (2022)
28. Loureiro, A.: Innovation and technology—the only answer for sustainable tourism growth. Worldwide Hospitality Tourism Themes **11** (2019)
29. Loureiro, S.M.C., Nascimento, J.: Shaping a view on the influence of technologies on sustainable tourism. Sustainability **13**(22) (2021)
30. Ma, Y., Tang, H., Ren, J.: Tourists' risk perceptions and pro-environmental behaviour during the pandemic: the roles of environmental concern and environmental ethics reflection. In: Current Issues in Tourism (2024)
31. McKinsey: Mckinsey & Company (2024). [Online]. Available at: https://www.mckinsey.com/featured-insights/mckinsey-explainers/what-is-the-internet-of-things
32. Nam, K., Dutt, C.S., Chathoth, P., Khan, M.S.: Blockchain technology for smart city and smart tourism: latest trends and challenges. Asia Pacific J. Tourism Res. (2021)
33. Otowicz, M.H., Macedo, M., Biz, A.A.: Dimensions of smart tourism and its levels: an integrative literature review. J. Smart Tourism (2022)
34. Pencarelli, T.: The digital revolution in the travel and tourism industry. Inf. Technol. Tourism **22**, 455–476 (2020)
35. Pierce, A.: Financing Profile: Small and Medium-Sized Enterprises in Tourism Industries (2016)
36. Rane, N., Choudhary, S., Rane, J.: Sustainable Tourism Development Using Leading-edge Artificial Intelligence (AI), Blockchain, Internet of Things (IoT), Augmented Reality (AR) and Virtual Reality (VR) Technologies. SSRN (2023)

37. Rashideh, W.: Blockchain technology framework: current and future perspectives for the tourism industry. In: Tourism Management (2020)
38. Raturi, G.: Medium (2023). [Online] Available at: https://medium.com/coinmonks/the-rise-of-metaverse-in-tourism-complete-guide-c8b797254227
39. Rosário, A.T., Dias, J.C.: Exploring the landscape of smart tourism: a systematic bibliometric review of the literature of the internet of things. In: Innovations in Sustainable Tourism: Shaping the Future of Destinations (2024)
40. Saurabh, K., Saxena, A.: Blockchain Technology Concept and Applications, 1st edn. Wiley (2020)
41. Sheehan, L., Vargas-Sánchez, A., Presenza, A., Abbate, T.: The use of intelligence in tourism destination management: an emerging role for DMOs. Int. J. Tourism Res. **18**(6) (2016)
42. Shone, M.C., Simmons, D.G., Dalziel, P.: Evolving roles for local government in tourism development: a political economy perspective. J. Sustain. Tourism **24**(12) (2016)
43. Sustacha, I., Pino, J.F.B., Valle, E.D.: The role of technology in enhancing the tourism experience in smart destinations: a meta-analysis. J. Destination Mark. Manage. **30** (2023)
44. Tham, A., Sigala, M.: Road block(chain): bit(coin)s for tourism sustainable development goals? J. Hospitality Tourism Technol. (2020)
45. Thees, H., Erschbamer, G., Pechlaner, H.: The application of blockchain in tourism: use cases in the tourism value system. Eur. J. Tourism Res. (2020)
46. Treiblmaier, H., Leung, D., Kwok, A., Tham, A.: Cryptocurrency adoption in travel and tourism–an exploratory study of Asia Pacific travellers. In: Current Issues in Tourism (2021)
47. Tsai, L.-F., et al.: Application of Kano model to discuss tourist demands on service quality in recreational travel. IJCSNS Int. J. Comput. Sci. Netw. Security (2017)
48. Tyan, I., Yague, M., Plaza, A.G.: Blockchain technology for smart tourism destinations. Sustainability **12**(22) (2020)
49. UNCTAD: Global Report on Blockchain and its Implications on Trade Facilitation Performance (2023)
50. UNWTO: Glossary of Tourism Terms (2015)
51. Weaver, D.: Sustainable Tourism. Routledge, s.l. (2007)
52. WTTC: Travel & Tourism Economic Impact Research (EIR) (2021)
53. Wu, Y., et al.: P2P energy trading: Blockchain-enabled P2P energy society with multi-scale flexibility services. In: Energy Reports (2022)
54. Zvaigzne, A., et al.: Smart tourism: the role and synergies of stakeholders. Worldwide Hospitality Tourism Themes **15**(5) (2023)

Chapter 9
Disruptive Technologies Reshaping Supply Chain Management for the Future: AI Integration, IoT, and Blockchain

Usha Badhera, Pooja S. Kushwaha, and Srikant Gupta

Abstract This study examines how disruptive technologies such as Artificial Intelligence, Blockchain, and the Internet of Things are transforming supply chain management and accelerating the shift toward Industry 4.0. These technologies are redefining operations by improving transparency, efficiency, and responsiveness while enabling innovative business models. Blockchain offers decentralized and tamper-proof ledgers that enhance trust and traceability across supply networks. Artificial Intelligence supports advanced predictive analytics, demand forecasting, and process automation, reducing costs and strengthening decision-making. The Internet of Things provides real-time visibility through data sharing and monitoring, which enhances logistics and enables agile supply chain operations. Although these technologies present remarkable opportunities, they also pose challenges such as high implementation costs, data integration complexity, scalability concerns, and cybersecurity risks. Even with these barriers, their benefits in terms of cost optimization, operational resilience, and improved customer satisfaction underscore their growing importance. This study emphasizes that organizations should view disruptive technologies not only as tools for efficiency but also as enablers of sustainable growth, competitive advantage, and long-term value creation in global markets.

Keywords Disruptive technologies · Supply chain management · Artificial Intelligence · Blockchain · Internet of Things

U. Badhera
Department of Business Analytics, Jaipuria Institute of Management, Jaipur, Rajasthan, India
e-mail: usha.badhera@jaipuria.ac.in

P. S. Kushwaha
Department of Business Analytics, Jaipuria Institute of Management, Indore, Madhya Pradesh, India
e-mail: pooja.singh@jaipuria.ac.in

S. Gupta (✉)
Department of Operations and Decision Sciences, Jaipuria Institute of Management, Jaipur, Rajasthan, India
e-mail: operations.srikant@hotmail.com

P. M. Jeyanthi (ed.), *Distributed Ledger Technology in Communication*, Signals and Communication Technology, https://doi.org/10.1007/978-3-032-07682-3_9

9.1 Introduction

Disruptive technologies redefine the landscape of supply chain management (SCM) by fundamentally altering how products or services are created, delivered, and consumed. The integration of disruptive technologies like Artificial Intelligence (AI) and robotic process automation integration, Internet of Things (IoT), and Blockchain (BC) innovations. AI means algorithms that mimic human cognitive functions; these process vast and complex datasets, discerning intricate patterns and trends with precision. BC technology represents a decentralized, distributed ledger system that ensures transparency, security, and immutability [29]. IoT is a vast network of interconnected and intelligent objects. These objects can seamlessly exchange information and services over the Internet, enabling the recording, monitoring, and optimization of their activities in real time. These disruptive technologies within SCM have not only catalyzed the emergence of innovative business models but also spearheaded the creation of new market opportunities and profound restructuring within existing ones. The core of these disruptive technologies is creating alternative values, methods, or cost structures that challenge traditional norms [22, 30]. By introducing solutions that are cheaper to produce, more straightforward, better performing, and more convenient to use, disruptive technologies outperform existing solutions and quickly surpass established markets. This sudden transformation reshapes industry dynamics, posing challenges to existing players while opening new avenues for growth and innovation. The impact of disruptive technologies extends beyond the business realm, affecting growth, employment, and inequality across different labor skills. This transformative force reshapes markets and influences labor, impacting workers' incomes and income distribution [20]. Despite the disruptions these technologies bring, it holds the potential to benefit consumers by providing access to cheaper and more accessible goods or services. In SCM, disruptive technologies are revolutionizing traditional practices and processes. By optimizing efficiency, enhancing transparency, and streamlining operations, these innovations are making the means for highly agile, responsive, and competitive SCM ecosystem. From robotics process automation and AI-driven predictive analytics to BC-enabled transparency and IoT-powered real-time monitoring, disruptive technologies are restructuring SCM for the future, driving value creation and sustainable growth.

9.2 Overview of Industry 4.0 and Its Influence on Supply Chains

Introducing Industry 4.0, also referred to as the fourth industrial revolution, is marked by the integration of innovative technology into manufacturing environments globally. Thanks to advancements in technology, this era saw a shift toward decentralizing corporate operations. Industry 4.0 signifies the significant technological advancements that have recently transformed the industrial sector globally. Simply put,

Fig. 9.1 Supply chain framework with integral role of logistics and transportation processes

it's a method that uses new technology to update traditional production processes, moving from machine-focused manufacturing to digital manufacturing [17]. Within industrial processes, there is a significant shift toward increased interconnection and intelligence. With Industry 4.0 technologies, businesses can now benefit from intelligent process management and embrace new paradigms in industrial management. By combining information and communication networks, these technologies enable manufacturing processes to be autonomous and dynamic [28].

Illustrated in Fig. 9.1 is the SCM structure, emphasizing the crucial role of logistics and transportation activities. According to [18], these procedures help connect all establishments involved in the SCM, ensuring efficiency and resource availability for smooth operations. The main focus of transportation procedures is facilitating the movement of goods and information among the various entities within the SCM.

9.3 Prominence of Adopting Disruptive Technologies in Supply Chain Management

Embracing new technologies is crucial for the prosperity of businesses, as lagging behind in their adoption can lead to major obstacles. Disruptive technologies are significantly impacting the field of SCM, transforming conventional processes and enhancing their effectiveness. The supply chain industry stands to gain from various technologies such as AI, robotics, the IoT, and BC. The IoT technology improves operational efficiency across the SCM by enhancing visibility, optimizing scheduling, and boosting customer satisfaction. The BC technology guarantees transparency, security, and transaction efficiency, leading to cost and risk reduction. Algorithms driven by AI improve SCM efficiency through optimizing inventory control, demand forecasting, and transportation network design. In addition, the integration of AI with robotics is leading to a transformation in the manufacturing process by enhancing flexibility, safety, and productivity. Organizations can successfully navigate through disruptions by revamping their supply chains and gaining a competitive advantage through the strategic utilization of disruptive technology. It is crucial for businesses to incorporate disruptive technologies into their SCM strategies to succeed in constantly changing market conditions, even though adoption rates may differ.

9.4 BC in Supply Chain Management

BC is an innovative technology that can lead to significant changes in the business environment and continuously escalate its application avenues. It has been recognized to have applicability in supply chain processes. In the past few years, BC has been recognized as a silver bullet for organizations, and its contribution to cloud computing [7, 27] revolutionized the businesses thus gained lots of attention from many industries. Blockchain's Internet-based technology is renowned for validating, recording, and disseminating transactions in immutable encrypted ledgers that nobody can alter. The system was developed to facilitate transactions in Bitcoin, a digital money that functions independently from a central bank. The wave of digitalization has made possible the live tracking of everything. To address the growing challenges of supply chain security planning, particularly for large, extremely complex supply chains, the industry has seen a remarkable transition over the past two decades. The idea of Blockchain is to establish a secure environment where buyers and sellers may conduct encrypted business without the involvement of third parties. This suggests that it might completely alter supply networks. As compared to the traditional supply chain, the use of technology and other Industry 4.0-related devices has enhanced performance and operational excellence. It becomes an inventive, eco-friendly, and productive supply chain [29].

9.4.1 Challenges

BC is a costly, high-maintenance storage technology. It will be beneficial only if the incurred benefits balance its higher cost. Since its implementation is expensive, it should only be used to store vital information about partner interorganizational interactions where visibility is crucial and trust is low. BC applications can contain a wide range of assets, from business papers (such as purchase orders and invoices) to pictures of goods and objects that stand in for actual physical entities. BC demands massive amounts of data in order to duplicate images of large sizes across all nodes in a network. The storage cost consists of physical networking and processing costs. The BC decision is coupled with some decisions related to storage, processing, and networking, and there are some environmental concerns. The technology also increases energy consumption, which enhances the use of natural resources to generate this energy.

The application of BC is limited by scalability concerns due to the constantly growing number of transactions, large block sizes, lengthy response times, and expensive fees. The increasing number of users poses a daily challenge to the scalability of BC technology. Data security, storage, and its management are some of the crucial elements of data management in BCs, other than data privacy and confidentiality. For BC to be applied effectively and function well, supply chain systems would need to be appropriately involved and participatory. Global logistics partners, maintenance costs, operational effectiveness, data management tools, and IT support would all be necessary for this. Significant implementation challenges are presented by government regulations pertaining to cryptocurrencies, data warehousing, scalability, high-speed internet connection, and the need for vast processing capacity. Every transaction needs to be processed and verified by BC technology; yet, in developing and underdeveloped nations, infrastructure problems and other constraints may render the goals of BC adoption useless [15]. Other issues that require research are the flexibility, standardization, expandability, and synergy of various technologies.

9.4.2 Benefits

Researchers and industry practitioners are trying to implement it concerning its advantages. One of the avenues to utilize its benefits is to improve the supply chain [10]. It would embrace the SCM as privacy, security, smart contract [4], scalability [21], and the ability to minimize cost and maximize profit would be considered the most sorted characteristics of BC. It can turn down the cost of labor as less manpower but more skilled manpower [31] is required. In the longer run, an imbibed supply chain system can achieve energy efficiency [33], cost efficiency, and improved performance [13]. It introduces innovation and generates opportunities for traceability and control of the goods by making it possible to track the history of a product and also supports inspection at any stage, resulting in a more effective and efficient supply chain [19, 25]. It helps to create awareness for the end customers and provides all the relevant information about the material they are interested in purchasing [2]. Transaction costs and human errors can be reduced by using this technology, which reduces time and risk and mitigates the chances of hacking. By tracking the product, customers' worries are reduced [9].

9.5 AI in Supply Chain Management

AI is one of the disruptive technologies that define the fourth industrial revolution. The conventional line separating humans from machines is blurred by it. AI was created to produce thinking abilities similar to those of humans, with the capacity to make judgements based on human knowledge and intelligence. AI has demonstrated its ability to enhance various business scenarios across nearly all industries since the 1970s, and SCM is no exception. Global e-commerce is expanding quickly as a result of smartphone and internet usage. One of the main drivers of the expansion of the local and international supply chain is e-commerce. Effective SCM is critical at a time of rising supply risk, more uncertain demand, and fiercer competition. Professionals in the supply chain have investigated numerous approaches to handle and use information to improve business choices, realizing its growing importance for the success of the chain. AI, which has been around for decades but hasn't been extensively used in SCM, could be one of those methods. [14].

AI has shown to be beneficial for projection and forecasting. Companies spends lot of effort and time to maintain a balance between supply and demand. As a result, its supply chain and manufacturing need better forecasting. AI has the capacity to analyze, forecast, and process data [5]. It aids in the provision of precise and trustworthy demand forecasts, enabling companies to source as cheaply as possible for purchases and order fulfilment. This would lower the expenses for SCM, warehousing, and transportation. One promising approach to improving automation and productivity in SCM is the integration of robotic process automation (RPA) and AI. This technology automates manual tasks that have historically been performed by humans by combining machine learning, AI, and software. The algorithms and programming of robots can automate repetitive administrative tasks and improve SCM processing. This removes operational errors and duplicates repetitive manual judgements, especially those that are rule-based and include a lot of data in digital systems. Therefore, robot processing and automation becomes one of the most important digital transformation technologies for automating repetitive SCM activities. Furthermore, it leverages AI's decision-making powers in conjunction with other technologies to improve technical proficiency and unlock automation potential across various SCM domains, which leads to intelligent automation [24]. RPA automates repetitive tasks while leveraging AI for enhanced decision-making, leading to intelligent automation and improved process efficiency [11, 24].

9.5.1 Challenges

The usage of AI in SCM has challenges that businesses may encounter during AI implementations. The complexity of data integration is a major challenge faced by the organizations. Supply chains generate vast amounts of data from various sources, including sensors, IoT devices, and enterprise systems. Integrating data from all these sources is a difficult task that requires significant time and resources. There is a surge of skilled personnel with expertise in SCM and AI technologies. Implementing AI effectively requires a multidisciplinary approach involving professionals in data science, logistics, and operations management [16]. AI algorithms rely heavily on data quality and accuracy. Inaccurate or incomplete data can lead to incorrect predictions and impact decision-making [32]. Data quality requires robust data governance practices and investment in data cleansing and validation processes. Data privacy and security are also concerns for SCM as they contain sensitive information about customers, suppliers, and operations. Protecting this data from unauthorized access and breaches requires robust cybersecurity measures and following compliances and regulations as per the country.

Cultural barriers also exist in AI adoption within organizations. Some employees may be uncomfortable with new technologies like AI and consider it a threat to their jobs. Overcoming these barriers requires effective change management strategies, including education and training programs to upskill employees and foster a culture of innovation and collaboration [23]. AI offers significant potential to transform SCM, businesses must address these challenges to successfully adopt and integrate AI technologies into SCM operations.

9.5.2 Benefits

AI offers numerous benefits in SCM, which helps in revolutionizing the business's operations. The significant contribution is in optimizing the inventory with the use of predictive analytics. Analyzing historical data and considering current market trends, AI helps forecast the demand accurately and helps businesses maintain their inventory [6]. This would help reduce inventory costs and, at the same time, ensure that the products are available as per the customers' demand, which helps enhance customer satisfaction.

AI also helps in improving the logistics and transportation of goods. AI-powered transportation systems can optimize routes, plan the vehicle load, and design delivery schedules, which leads to reduced transportation costs and faster delivery times. AI helps in the real-time tracking of shipments [8]. AI also plays a crucial role in enhancing production processes. Manufacturers can optimize production schedules, minimize downtime, and improve equipment effectiveness by leveraging machine learning algorithms. AI enables supplier management through data-driven insights. By analyzing supplier performance and market trends, businesses can identify opportunities for cost savings and negotiation terms that would help mitigate supply chain risks. By harnessing AI technologies, businesses can gain a competitive edge, reduce costs, and better meet customer demands [12].

9.6 IoT in Supply Chain Management

Traditional methods of SCM are plagued by problems like as overstocking, delivery delays, and stockouts; however, these issues may be resolved with the use of IoT technology. The IoT makes it possible to communicate and share data in a seamless manner at any time, in any location, and through a variety of channels. Businesses in the supply chain may reduce the costs associated with knowledge acquisition by employing intelligent IoT devices, which also improves the supply chain's intelligence and efficiency. The research that was carried out by Abdel-Basset and his colleagues in 2018 explains that supply chains may reduce the costs associated with knowledge acquisition by employing intelligent IoT devices, which also improves the supply chain's intelligence and efficiency. In popular parlance, the term IoT refers to the gadgets that are outfitted with sensors and actuators. In order to do this, it is necessary to gather, analyze, and share data across the many components, applications, and platforms, as well as with one another. By allowing for the integration of

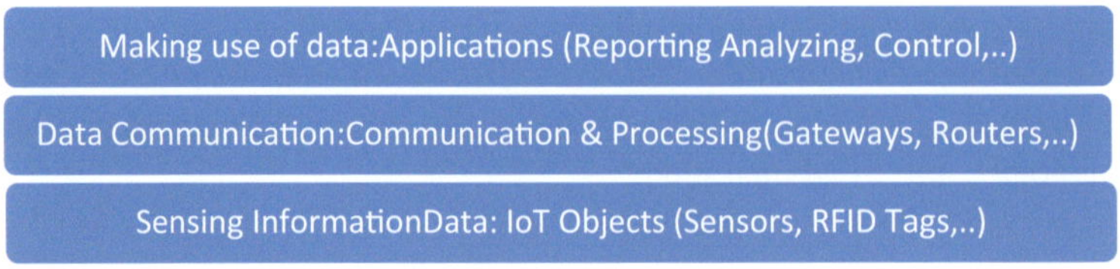

Fig. 9.2 Basic components of IoT

digital systems and physical items, the IoT makes it possible to construct intelligent applications and structures. This is made possible by the connection that it provides. It is possible for several intelligent physical things to communicate with one another and collaborate in order to achieve a shared objective, as seen in Fig. 9.2. Examples of devices that are frequently utilized in a variety of applications include wearables, sensors, actuators, telephones, automobiles, computers, RFID, and many others. By utilizing one-of-a-kind addressing systems and communication protocols that are standardized, these gadgets are able to collaborate with other smart devices that are strategically located nearby.

There have been a lot of new applications in the field of logistics and supply chains in recent years, and the IoT has the potential to have a significant impact on supply chains in the future. SCM is the process of meticulously coordinating all of the pieces that make up the supply chain for goods and services. The IoT sector plays a significant role in improving modern SCM. By digitalizing and automating the activities involved in SCM, IoT technology offers the potential to reduce operational costs and improve operational efficiencies. The proliferation of IoT devices has resulted in a considerable evolution of supply chains. For the purpose of tracking and monitoring shipments, these gadgets make use of real-time monitoring technologies such as GPS. [3] The proliferation and growth of the IoT has the potential to completely transform the supply chains of today.

IoT revolutionizes SCM with its multifaceted capabilities. Using IoT sensors, SCM gains real-time insights through machine-readable data, enriching decision-making processes. Furthermore, IoT fosters seamless communication and collaboration by interconnecting smart objects, assets, and business entities via the Internet and IT systems. IoT-driven SCM optimizes decision-making across the entire lifecycle through business intelligence and smart analytics. Automation is a key benefit, as IoT streamlines SCM processes, reducing manual efforts, enhancing speed, and ensuring reliability. Integrated IoT-enabled SCM forms a cohesive network, facilitating collaboration and information sharing across various stages. Lastly, IoT drives innovation in SCM, introducing transformative solutions that cater to modern needs and enable adaptation to evolving market demands. The stages of RFID-IoT technologies in supply chain management are shown in Fig. 9.3.

Fig. 9.3 Stages of RFID-IoT technologies in SCM [26]

9.6.1 Challenges

Implementing IoT technologies in SCM introduces several challenges that organizations must navigate to leverage these advancements' benefits fully. Fluctuating consumer demand presents a significant hurdle, as varied preferences and technological advancements contribute to unpredictable demand patterns, complicating product supply and demand forecasting efforts. Moreover, integrating physical and digital systems into IoT-based SCM heightens cybersecurity risks, as it increases vulnerabilities across organizational and operational domains due to insufficient standardization and data-sharing protocols.

Additionally, organizations face high investment costs when transitioning to IoT-based SCM, requiring substantial investments in infrastructure such as network devices, sensors, and data processing systems. Furthermore, Inadequate Network Infrastructure poses a challenge, particularly in geographically dispersed SCM processes, where gaps hinder real-time data flow and traceability. Overcoming these challenges demands strategic planning, investment, and collaboration to ensure IoT technologies' effective implementation and utilization in SCM.

9.6.2 Benefits

Integrating IoT technologies into SCM offers several key applications and benefits. QR Code and RFID Integration optimizes inventory management by efficiently tracking products, enhancing visibility, and ensuring accuracy throughout the supply chain. Smart Tags and NFC Connectivity enable real-time monitoring and secure data exchange, ensuring product quality and authenticity at every stage. IoT for enhanced monitoring provides real-time visibility into SCM processes, facilitating proactive monitoring and tracking of goods, thereby enhancing transparency and responsiveness across the supply chain. These applications empower organizations to streamline operations, improve efficiency, and meet evolving consumer demands in today's dynamic market landscape.

Sensors and connected devices facilitate data-driven decision-making in SCM, which gathers comprehensive data on diverse supply chain parameters. This wealth of information empowers organizations to make informed decisions and fine-tune processes for heightened efficiency. IoT applications in logistics are also reshaping the landscape by facilitating real-time asset monitoring and route optimization and bolstering overall supply chain agility. These advancements are particularly impactful in revolutionizing warehousing and transportation processes and improving operational efficiency and responsiveness to dynamic market demands [1].

9.7 Way Forward

BC is a storage medium known for its high cost and significant amount of additional work required. When considering privacy, security, smart contracts, scalability, and the ability to reduce costs while increasing profits, it would be beneficial to incorporate them into SCM. It is conceivable that this could help reduce labor costs by necessitating a smaller workforce with more specialized skills. BC technology is only practical when its benefits exceed its higher cost. One of the primary challenges that businesses face is the complexity of data integration. AI could help optimize inventory by using predictive analytics. An accurate estimation of demand and the maintenance of company inventories can be achieved by analyzing historical data and current market trends. Discussing the migration to infrastructure like network devices, sensors, and data processing systems, business process management related to the IoT requires significant investment. Inadequate standardization and norms for data exchange could lead to an increase in vulnerabilities across operational and organizational domains. One can enhance inventory management by incorporating the IoT to track items effectively, improve visibility, and ensure accuracy throughout the supply chain. All these innovative technologies will play a role in transforming SCM. To effectively implement these technologies, organizations must make informed decisions considering their culture and the current market demand.

References

1. Abdel-Basset, M., Manogaran, G., Mohamed, M.: Internet of Things (IoT) and its impact on supply chain: a framework for building smart, secure and efficient systems. Futur. Gener. Comput. Syst. **86**(9), 614–628 (2018)
2. Bechini, A., Cimino, M.G., Marcelloni, F., Tomasi, A.: Patterns and technologies for enabling supply chain traceability through collaborative e-business. Inf. Softw. Technol. **50**(4), 342–359 (2008)

3. Ben-Daya, M., Hassini, E., Bahroun, Z.: Internet of things and supply chain management: a literature review. Int. J. Prod. Res. **57**(15–16), 4719–4742 (2019)
4. Christidis, K., Devetsikiotis, M.: Blockchains and smart contracts for the internet of things. IEEE Access **4**, 2292–2303 (2016)
5. Dash, R., McMurtrey, M., Rebman, C., Kar, U.K.: Application of artificial intelligence in automation of supply chain management. J. Strateg. Innov. Sustain. **14**(3), 43–53 (2019)
6. Fosso Wamba, S., Queiroz, M.M., Guthrie, C., Braganza, A.: Industry experiences of artificial intelligence (AI): benefits and challenges in operations and supply chain management. Prod. Plann. Control **33**(16), 1493–1497 (2022)
7. Giannakis, M., Spanaki, K., Dubey, R.: A cloud-based supply chain management system: effects on supply chain responsiveness. J. Enterp. Inf. Manag. **32**(4), 585–607 (2019)
8. Helo, P., Hao, Y.: Artificial intelligence in operations management and supply chain management: an exploratory case study. Prod. Plann. Control **33**(16), 1573–1590 (2022)
9. Iansiti, M., Lakhani, K.R.: The truth about blockchain. Harv. Bus. Rev. **95**(1), 118–127 (2017)
10. Kleeff, J., Korc, M., Apte, M., La Vecchia, C., Johnson, C.D., Biankin, A.V., Neoptolemos, J.P., et al.: Pancreatic cancer. Nat. Rev. Dis. Primers. **2**(1), 1–22 (2016)
11. Kudlak, L.: Don't Underestimate the Power of Robotic Process Automation. Will the Age of Ultron Come to Our World? (2019). https://medium.com/tech4planet/dont-underestimate-the-powerofrobotic-process-automation-8ffb8262d62f
12. Lebhar, I., Dadda, A., Ezzine, L.: Artificial Intelligence applications in the global supply chain: benefits and challenges. In: International Conference on Advanced Intelligent Systems for Sustainable Development, pp. 282–295. Springer Nature Switzerland, Cham (2022)
13. Makridakis, S., Christodoulou, K.: Blockchain: current challenges and prospects/applications. Future Internet **11**(12), 258 (2019)
14. Min, H.: Artificial intelligence in supply chain management: theory and applications. Int. J. Log. Res. Appl. **13**(1), 13–39 (2010)
15. Min, H.: Blockchain technology for enhancing supply chain resilience. Bus. Horiz. **62**(1), 35–45 (2019)
16. Nozari, H., Szmelter-Jarosz, A., Ghahremani-Nahr, J.: Analysis of the challenges of artificial intelligence of things (AIoT) for the smart supply chain (case study: FMCG industries). Sensors **22**(8), 2931 (2022)
17. Oztemel, E., Gursev, S.: Literature review of Industry 4.0 and related technologies. J. Intell. Manuf. 1–56 (2018)
18. Pečený, L., Meško, P., Kampf, R., Gašparík, J.: Optimisation in transport and logistic processes. Transp. Res. Proc. 15–22 (2020)
19. Rayes, A., Salam, S., Davila, C., Tarnow, J.: The Blockchain in IoT. Internet of Things From Hype to Reality: The Road to Digitization, pp. 269–296(2019)
20. Reichstein, C.: Strategic IT management: how companies can benefit from an increasing IT influence. J. Enterp. Inf. Manag. **32**(2), 251–273 (2019)
21. Scherer, M.: Performance and Scalability of Blockchain Networks and Smart Contracts (2017)
22. Sharma, P.K., Kumar, N., Park, J.H.: Blockchain technology toward green IoT: opportunities and challenges. IEEE Netw. **34**(4), 263–269 (2020)
23. Sharma, R., Shishodia, A., Gunasekaran, A., Min, H., Munim, Z.H.: The role of artificial intelligence in supply chain management: mapping the territory. Int. J. Prod. Res. **60**(24), 7527–7550 (2022)
24. Siderska, J., Aunimo, L., Süße, T., von Stamm, J., Kedziora, D., Aini, S.N.B.M.: Towards Intelligent Automation (IA): literature review on the evolution of Robotic Process Automation (RPA), its challenges, and future trends. Eng. Manage. Prod. Services **15**(4), 90–103 (2023)
25. Steiner, J., Baker, J., Wood, G., Meiklejohn, S.: Blockchain: The Solution for Transparency in Product Supply Chains (2015). Available at: provenance.org/whitepaper
26. Tan, W.C., Sidhu, M.S.: Review of RFID and IoT integration in supply chain management. Oper. Res. Perspect. **9**, 100229 (2022)
27. Taylor, P.J., Dargahi, T., Dehghantanha, A., Parizi, R.M., Choo, K.K.R.: A systematic literature review of blockchain cyber security. Digital Commun. Netw. **6**(2), 147–156 (2020)

28. Tortorella, G.L., Fettermann, D.: Implementation of industry 4.0 and lean production in Brazilian manufacturing companies. Int. J. Prod. Res. **56**(8), 2975–2987 (2017)
29. Upadhyay, A., Mukhuty, S., Kumar, V., Kazancoglu, Y.: Blockchain technology and the circular economy: implications for sustainability and social responsibility. J. Cleaner Prod. 126130 (2021)
30. Wang, Y., Chen, C.H., Zghari-Sales, A.: Designing a blockchain enabled supply chain. Int. J. Prod. Res. **59**(5), 1450–1475 (2021)
31. Yadav, S., Singh, S.P.: Blockchain critical success factors for sustainable supply chain. Resources Conserv. Recycling **152**, 104505 (2020)
32. Zeng, X., Yi, J.: Analysis of the impact of big data and artificial intelligence technology on supply chain management. Symmetry **15**(9), 1801 (2023)
33. Zhang, J., Zhong, S., Wang, T., Chao, H.C., Wang, J.: Blockchain-based systems and applications: a survey. J. Internet Technol. **21**(1), 1–14 (2020)

Chapter 10
Supply Chain Tracking and Management with Blockchain

Violeta Cvetkoska, Ljupcho Eftimov, Kiril Postolov, and Bojan Kitanovikj

Abstract Blockchain technology is increasingly recognized as a tool for enhancing supply chain tracking and performance management. This chapter examines how blockchain, as a distributed and immutable ledger system, can address longstanding supply chain challenges, including performance management, fragmented data, distrust among partners, and reliance on intermediaries. The literature and recent implementations across industries help us assess the blockchain's impact on transparency, traceability, and operational efficiency in supply chains. The chapter introduces the fundamentals of integrating blockchain into supply chain processes and examines its role as a performance management tool enabling real-time, trustworthy metrics of supply chain activities. Four in-depth case studies from diverse sectors (food retail, pharmaceuticals, logistics, and luxury goods) illustrate the current applications of blockchain in supply chain tracking and management. A comparative analysis framework is developed to evaluate each case along dimensions of strengths, opportunities for growth, and specific blockchain use cases. We also conceptualize how blockchain networks interact with supply chain layers and data flows to reduce intermediaries and improve transparency. Furthermore, emerging trends are identified, including blockchain's convergence with Internet of Things sensors and smart contracts, and its potential to drive multi-industry supply chain innovation by fostering trust, efficiency, and data-driven performance management in both developed and developing countries.

V. Cvetkoska · L. Eftimov · K. Postolov · B. Kitanovikj (✉)
Ss. Cyril and Methodius University in Skopje, Faculty of Economics—Skopje, Skopje, North Macedonia
e-mail: bojan.kitanovikj@eccf.ukim.edu.mk

V. Cvetkoska
e-mail: vcvetkoska@eccf.ukim.edu.mk

L. Eftimov
e-mail: ljupco.eftimov@eccf.ukim.edu.mk

K. Postolov
e-mail: kiril.postolov@eccf.ukim.edu.mk

P. M. Jeyanthi (ed.), *Distributed Ledger Technology in Communication*, Signals and Communication Technology, https://doi.org/10.1007/978-3-032-07682-3_10

Keywords Supply chain management · Blockchain · Performance management · Supply chain tracking

10.1 Introduction

Blockchain has become the center of attention in multiple industries, including finance and banking, healthcare, information technology, and the public sector [29]. This rise in popularity can be explained by the fact that these and other industries heavily depend on intermediaries. With blockchain, the operations of these industries can become decentralized, eliminating the necessity to establish verification systems, while maintaining reliability and information security in the processes [20]. In short, this technology presents the opportunity to host an extensive accounting and information ledger for recording transactions [22]. As such, it has become a powerful tool for tracking and performance measurement in this context, considering that all supply chain-related information is transparent, authentic, traceable, and secure.

Academics, practitioners, and industry insiders consider blockchain to be one of the most prominent technologies with the potential to disrupt modern working and modern supply chain management [30]. Hence, the application of this technology in the supply chain is still in the initial stages, which is one of the primary motivations behind this book chapter. The opportunities, the present situation, and the potential future uses in supply chain management can remodel and reshape supply and logistics [54]. Additionally, from the point of view of organizational sciences, little is still known about the potential to use blockchain for performance management. It has been shown to alter management procedures through secure record keeping, credential verification, supplier management, payroll management, talent management, decentralized organizational units, and more [27].

Conceptually, the objective of this research is: (1) to reintroduce blockchain into the supply chain management literature as an avenue for efficient, automated, secure, and transparent performance measurement and tracking in a variety of industries, as well as, (2) to analyze the current applications and areas where blockchain can be embedded in managing supply chains, and (3) to propose future blockchain applications and recommendations for business leaders and supply chain and logistics personnel.

Based on a thorough review of the literature and analysis of selected use cases, the book chapter proposes a blockchain-based framework along with the use of other affiliated technologies for supply chain management, which brings to the surface the feasible benefits of this technology. Through the illustrative design of this process, we propose an achievable use case of a more efficient performance management system within supply chain management, which can be tested and applied in other real-world environments.

The book chapter is structured as follows. First, the abstract presents a concise overview of the research. Next, the introduction sets the stage by outlining the context and significance of blockchain in supply chain management. In the related work segment, existing literature is reviewed to establish the study's foundation. The book chapter then offers an introduction to blockchain integration in supply chain management, highlighting fundamental concepts, key benefits, as well as alternatives for how blockchain facilitates metrics and evaluation processes. Following this, the part related to applications of blockchain in supply chain tracking and management demonstrates real-world use cases and implementations. The penultimate section concerning the future of blockchain in supply chain tracking and management evaluates emerging trends and prospective advancements, eventually ending with some concluding remarks.

The findings noted in this book chapter carry several implications for both research and practice. On the research front, they highlight the need for further investigation into how decentralized technologies can reshape performance management, extending beyond mere transaction tracking to broader organizational processes. Practically, blockchain's capacity to enhance security, transparency, and authenticity in supply chain information flows can radically reduce intermediary reliance, thus streamlining operations and advancing cost efficiency [73]. As such, this promises to transform the work of business leaders and supply chain and logistics managers, signaling a paradigm shift in how organizations design and measure performance.

10.2 Related Work

Global supply chains are complex networks involving multiple organizations, from raw material suppliers to manufacturers, distributors, and retailers. Often, managing and monitoring such chains is seen as challenging due to siloed information systems, limited transparency, and the need for trust among parties [75]. Blockchain technology, a distributed digital ledger known for underpinning cryptocurrencies, has emerged as a promising solution to many of these issues [34]. By enabling a decentralized, tamper-proof record of transactions and data exchanges, blockchain can greatly improve supply chain operations. Specifically, researchers note that blockchain can enhance transparency and traceability, improve coordination among partners, increase the speed and efficiency of deliveries, and reduce the need for costly intermediaries or paperwork [36, 37].

Blockchain in supply chain management has attracted considerable scholarly attention in recent years. Early works established the foundational view that blockchain could support key supply chain objectives such as cost reduction, quality improvement, speed, dependability, risk mitigation, and sustainability [51]. For instance, Kshetri [40] presented some of the first case evidence linking blockchain use to greater transparency and accountability in supply chains. Around the same time, Saberi et al. [57] critically examined how blockchain and smart contracts might enhance sustainable supply chain management and identified core benefits, while also discussing barriers to adoption between organizations, in organizations, related to technology, the environment, and more.

Since then, numerous literature reviews and empirical studies have been published, reflecting the maturation of the field. Several systematic literature reviews have mapped the landscape of blockchain applications in supply chain management. Shoaib et al. [62] conducted a bibliometric analysis, concluding that blockchain has been widely explored as a tool to improve transparency, information sharing, process efficiency, and traceability in supply chains. This highlights that research interest grew exponentially after 2017, marked by recent technological, regulatory, and social advancements. Khan et al. [38] also confirmed that blockchain is gaining high utility in supply chain management and is constantly evolving with technological advances. In recent instances, it has been found to integrate with other emerging technologies such as the Internet of Things (IoT) to enhance data capture and analytics [8].

A strand of research has focused on supply chain performance results related to blockchain adoption. Namely, dimensions such as agility, alignment, and adaptability have been discussed in the context of the marriage between blockchain and supply chain management. Findings suggest that real-time information sharing via blockchain can improve responsiveness to demand changes, while the immutable unified ledger enhances trust among partners, indirectly boosting performance [50]. Blockchain's impact on supply chain resilience and robustness has been explored, too, indicating that the technology can facilitate better risk management and continuity planning by providing end-to-end visibility [70].

On the other hand, scholars also caution that performance gains are contingent on broad adoption, meaning that a blockchain network's value increases with more participants, whereas partial implementations (limited to a few nodes in the chain) yield only limited improvements [48].

Another important body of related work examines barriers and challenges to implementing blockchain in supply chains. Commonly cited challenges include scalability and transaction speed issues with some blockchain platforms, regulatory and legal uncertainties (especially around smart contracts and data sharing), data privacy concerns in sharing information on a ledger, and the need for standardization across industry players [42, 57, 73]. Organizational barriers such as lack of top management support or limited blockchain expertise also emerge [72].

To address these, researchers have proposed frameworks for blockchain adoption, often emphasizing an incremental approach to start with pilot projects in a controlled scope, ensure clarity on the governance of the blockchain network, and build consortiums or industry alliances to drive standards [48]. In support of this, Casino et al. [17] developed a taxonomy of blockchain platforms for different supply chain needs, guiding managers on whether to use public, private, or consortium blockchains depending on required throughput and trust levels.

Recent studies highlight that blockchain's applicability spans multiple industries and sectors, such as food and agriculture, healthcare, manufacturing, and infrastructure supply chains [58]. This multi-industry relevance stems from common supply chain pain points that blockchain can address. In food supply chains, blockchain is employed to track products from farm to table, assuring quality and safety [56]. In pharmaceuticals, it helps verify drug authenticity and proper handling conditions

[3]. In manufacturing and retail, blockchain improves the traceability of components and ethical sourcing of materials [43]. Across these domains, blockchain provides a shared source of truth that all stakeholders can trust, thereby reducing the reliance on third-party auditors or paper-based certification processes. By recording every transaction or handover on an immutable ledger, participants gain the ability to instantly trace products and verify data, which was previously either impossible or extremely time-consuming with legacy systems [40].

Another driving force for blockchain adoption is the increasing emphasis on supply chain sustainability and compliance. Regulatory pressures like food safety regulations or conflict minerals reporting requirements and consumer demand for ethically sourced products have made transparency a critical performance metric for supply chains [36]. So, blockchain's security and immutability ensure that once data (e.g., a batch's origin, timestamps, certifications) is on the ledger, it cannot be altered, thereby providing reliable evidence for compliance and sustainability claims. Scholars have pointed out that blockchain's decentralized consensus can build trust in shared supply chain information, allowing parties to rely on data without a central gatekeeper [4]. This has implications for performance improvement as companies can more easily monitor supplier performance, trace bottlenecks, and react to issues in near-real-time.

With all that being said, the consensus in recent studies is that while blockchain is not a cure-all solution for all supply chain problems, it offers a novel infrastructure for secure, decentralized information management that aligns well with the multi-actor nature of supply chains.

10.3 Introduction to Blockchain Integration in Supply Chain Management

Integrating blockchain into supply chain management requires understanding both the technology and the operational context of supply chains. Blockchain, at its core, is a distributed ledger where transactions or events are recorded in chained blocks, secured by cryptographic hashes [34]. These principles can be applied in a supply chain setting, where a blockchain can serve as a shared database that records key events and transactions across the lifecycle of a product [50]. For example, when raw materials are harvested when a product is manufactured, when it ships, and when it arrives at a distribution center or retail store. Unlike traditional databases controlled by a single entity, a blockchain network is typically shared among multiple organizations or nodes, each of which holds a copy of the ledger and participates in verifying new entries via a consensus mechanism [34].

There are several architectural approaches to implementing blockchain in supply chains. One common approach is the use of permissioned or consortium blockchains, where only authorized supply chain partners can participate in the network [6]. Alternatively, some projects use public blockchains such as Ethereum with additional layers of encryption to protect sensitive data [46]. The choice of blockchain platforms and types in terms of public, private, or consortium depends on factors like the number of participants, trust level between them, throughput needs, and governance considerations [25]. In many cases, industry consortia or lead firms can establish blockchain networks and invite suppliers and logistics providers to join, creating an ecosystem around a single source of truth for transactions such as IBM's Food Trust platform in the food industry [63].

Integrating blockchain also involves linking physical goods to digital records on the ledger. This often requires complementary technologies such as IoT devices and sensors. For instance, Radio Frequency Identification (RFID) tags, QR codes, or serial numbers can be used to represent physical items in the digital ledger [14]. When a tagged item moves through a supply chain checkpoint (like a warehouse or port), an IoT sensor or an information system can automatically record an event on the blockchain. Illustratively, the notification can take the form of "Package X received at Warehouse Y at time Z." This IoT-blockchain integration ensures that data recorded on the blockchain is grounded in physical reality, and it minimizes manual data entry [14]. A practical example is temperature sensors in refrigerated trucks that report readings to a blockchain when transporting perishable goods, so all parties can verify the cold chain was unbroken.

Another key aspect of integration is the use of smart contracts, representing a self-executing code on the blockchain that can automate business processes [24]. In supply chains, smart contracts can be programmed to trigger actions when certain conditions are met. Illustratively, a smart contract could automatically release a payment to a supplier once a shipment is delivered and logged on the blockchain, or it could send an alert if a sensor reports out-of-threshold conditions like temperature above a safe level for medicines. As a result, smart contracts facilitate workflow automation and performance management by enforcing predefined rules without human intervention, reducing delays and opportunities for dispute [24]. However, designing smart contracts requires careful consideration of legal and business logic alignment, and they must be agreed upon by all relevant parties in the chain [39].

From an organizational perspective, integrating blockchain often means re-engineering some processes. Companies must decide which data to record on the blockchain and who will have access to it [60]. This typically involves mapping the supply chain processes to identify pain points such as a potential lack of visibility or manual paperwork that blockchain could address. To achieve this, a governance structure is needed to determine how participants will reach a consensus on adding new records. Additionally, standards for data formats and interoperability should be set so that systems of different companies can interface with the blockchain [12].

A critical consideration in integration is the scalability and performance of the blockchain itself. Supply chains generate a vast number of transactions. So, if each item scan or transfer is a blockchain transaction, the system must handle high volume. Early blockchain pilots sometimes struggled with throughput, but newer protocols and off-chain scaling techniques are improving capacity. In this regard, some solutions batch multiple events into one blockchain transaction or use layer-two networks.

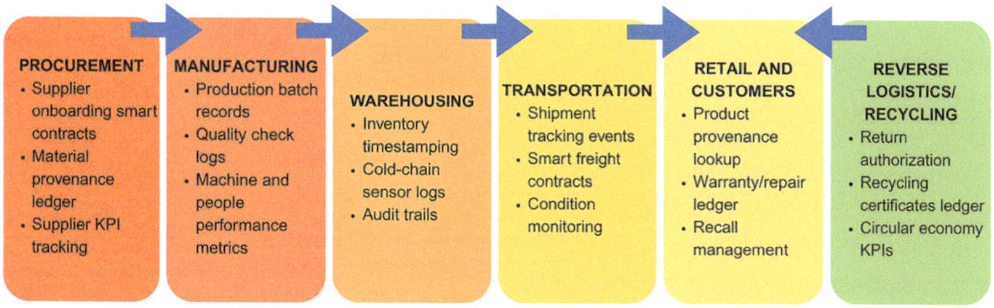

Fig. 10.1 Blockchain integration in supply chain management as a performance management tool. *Source* Adapted from [15, 26, 55]

Furthermore, Fig. 10.1 shows that integrating blockchain into supply chain management is a socio-technical endeavor as it combines technology deployment (blockchain network setup, possibly IoT integration, smart contract development) with process redesign and ecosystem governance. In this sense, this integration is viewed as a consistent loop where once the goods or services are delivered to the customer, the process of recycling and going backward into the process begins.

When done successfully, the result is a supply chain where each stakeholder can record and verify critical information in real-time, leading to a more transparent and responsive chain.

10.4 Blockchain as a Performance Management Tool

One of the less discussed but highly significant advantages of blockchain in supply chains is its ability to function as a performance management tool. In traditional supply chains, collecting performance data, such as on-time delivery rates, lead times, and temperature excursions, often involves reconciling records from different systems and trusting reports provided by various intermediaries. This process can be slow and prone to errors or even data manipulation [1]. Blockchain changes this paradigm by providing real-time, reliable data streams from the immutable ledger that all parties share [77]. Once key performance-related data are recorded on a blockchain, they become tamper-proof and easily accessible for analysis, enabling more accurate and timely measurement of supply chain performance metrics.

Blockchain facilitates valid and effective measurement of outcomes and performance of supply chain processes by ensuring data integrity [10]. For example, consider the metric of end-to-end order fulfillment time. In a blockchain-enabled supply chain, each milestone in terms of order placed, order shipped, customs cleared, delivered, and so on, could be logged as a transaction on the ledger with a timestamp. Because this ledger is immutable and shared, managers can trust the data and calculate the total lead time without worrying that any party has altered the timestamps to mask delays. In essence, blockchain provides a single source of truth for performance data. [40] observed that once input tracking data are on a blockchain ledger, they are immutable and thus can serve as a reliable basis for performance measurement. This immutability builds confidence that key performance indicators (KPIs) are based on actual events, not self-reported figures.

Another dimension is auditability. Supply chain performance often connects to compliance and quality audits such as ISO standards, food safety audits, and others, which are relevant to the industry [18]. Blockchain inherently keeps a full history of transactions, making it straightforward to audit the chain's activities over time. Instead of auditors spending weeks gathering documents from various sources, they could query the blockchain for relevant records with appropriate permissions [45]. For instance, a performance audit might require proof that refrigerated goods stayed within temperature limits throughout transit. If IoT sensors feed temperature data to the blockchain at regular intervals, an auditor can easily extract that history and verify compliance. Illustratively, in the agri-food sector, blockchain has been used to record every step of meat production, enabling auditors and even consumers to verify the product's journey and handling conditions, significantly streamlining the assurance of quality metrics [67].

Blockchain also reduces the need for intermediary oversight in performance tracking. Traditionally, companies might hire third-party inspectors or rely on central authorities to certify certain performance aspects like the case of a trade-finance intermediary verifying that goods have shipped before releasing payment. With blockchain, many of these verification tasks can be automated or disintermediated [32]. In performance management terms, this means companies can directly monitor partner performance such as delivery times, condition of goods, and similar, through blockchain records rather than waiting for periodic reports or certifications. This near-real-time visibility of performance indicators enables faster response to any deviations [44]. If a supplier is consistently late or a batch's quality data falls outside the acceptable range, the issue is apparent on the ledger immediately, prompting corrective action or renegotiation of terms if necessary.

Hence, smart contracts on blockchain can be used to automate performance management actions [77]. For instance, a procurement contract encoded as a smart contract might include a clause that automatically applies a penalty or discount if a delivery is late beyond a certain threshold, using the timestamp data on the blockchain as evidence. This kind of automated enforcement ensures that performance agreements are transparently monitored and executed without bias. It also incentivizes partners to meet their targets, knowing that the blockchain will record any lapse and trigger pre-agreed consequences [50].

Blockchain's role in performance management is also evident in quality control and traceability metrics. Consider the performance goal of reducing the rate of defective products reaching customers. Using blockchain, every component or ingredient of a product can be traced to its origin. If a defect is detected, one can quickly trace back through the ledger to identify the source in terms of the batch, supplier, or manufacturing lot. This rapid traceability is itself a performance metric as many companies now measure "traceability response time" [11].

Another aspect is continuous improvement. With comprehensive, trustworthy data, firms can perform deeper analysis of their supply chain processes. They can identify chronic bottlenecks, and benchmark partners across the supplier base more easily, fostering a performance-driven culture [2, 64]. It goes without saying that data privacy must be managed, considering that one supplier may not want competitors to see their metrics, but permissioned blockchains allow fine-grained access control so that aggregated or anonymized benchmarks can be derived [76].

Finally, blockchain's transparency fosters trust, which is a less quantifiable but crucial aspect of performance management in multi-firm networks [78]. When every participant knows that data on the ledger is correct and visible to all, there is an inherent incentive to uphold commitments. A popular saying in performance management is what gets recorded gets managed. This trust can improve collaborative performance measures, such as supply chain fill rate or perfect order rate because partners are more willing to share information and coordinate when they trust the data [19].

10.5 Blockchain Applications in Supply Chain Tracking and Management

In this section, we present four detailed case studies of companies that have implemented blockchain solutions in their supply chain operations. These cases, drawn from recent scientific and industry literature, span multiple industries, including food retail, pharmaceuticals, global shipping, and luxury goods, demonstrating the broad applicability of blockchain. For each case, we describe the specific blockchain application, the benefits and strengths observed, and the opportunities for further improvement. Namely, after the case descriptions, we provide a comparative analysis, summarizing strengths, growth opportunities, and the particular blockchain use in each instance. The cases were chosen to reflect real-world deployments instead of just proofs-of-concept in an effort for these cases to become a blueprint for scalability in other industries. This is why we chose four distinct and diverse industries that may not share many common aspects.

1. *Blockchain for food supply chain traceability (retail/food sector).* Walmart's blockchain initiative, developed with IBM on Hyperledger Fabric, began with 2016–2017 pilots tracking U.S. mangos and Chinese pork and logging every custody change from farm origin, batch, and lot codes, to processing details, temperature, and shipping data [40]. This was done onto an immutable shared ledger, and the result was a dramatic reduction in trace-back time from nearly seven days to 2.2 s, enabling pinpoint recalls that curb waste, safeguard consumer trust, and save millions otherwise lost in blanket withdrawals [65].

Another strength observed is improved supply chain transparency and trust among stakeholders. With all participants (farmers, processors, distributors, Walmart itself) entering data into a shared ledger, there is end-to-end visibility. Each party can see authenticated data about the product's journey. For example, store managers can verify which farm a crate of lettuce came from and when, and farmers can see that their products have safely arrived at stores. This transparency not only aids safety but also allows performance tracking such as monitoring how long products spend in transit or in warehouses (dwell time), which is important for quality management (e.g., ensuring fresh produce moves quickly). Walmart's blockchain trial indicated that such visibility could reduce administrative burdens as earlier, tracing shipments required checking siloed records and often contacting intermediaries [65].

Walmart also showed cost-saving implications. In crises involving contaminated food, the ability to do surgical recalls (thanks to blockchain-based traceability) means the company does not have to recall tons of safe products just in case, saving potentially millions of dollars [40]. It also helps regulators and health officials, for instance, if an E. coli outbreak is linked to a certain farm's produce, authorities can quickly confirm which supply chains are sourced from that farm. After successful pilots, Walmart proceeded to mandate that all its leafy greens suppliers integrate with its blockchain tracking system, aiming to improve the overall ecosystem's safety and performance. This rollout, as reported in 2019, was one of the first large-scale blockchain adoptions in retail supply chains [61].

Future growth lies in scaling thousands more products and suppliers, automating data capture through IoT sensors for richer real-time inputs, and achieving interoperability with other retailers' or industry-wide platforms so traceability persists beyond Walmart's network. Yet even in its current form, it is evident that blockchain's capacity to speed and secure supply chain tracking while reducing manual verification and intermediary dependence can bring benefits to the food and retail industry.

2. *Blockchain for drug shipment monitoring (pharmaceutical/healthcare sector).* Gemalto's blockchain-enabled cold-chain pilot for temperature-sensitive vaccines and medicines illustrates how an immutable, shared ledger can safeguard pharmaceutical integrity end-to-end [23]. Hence, digital data loggers embedded in shipping containers continuously captured temperature, time-stamped each reading, and wrote it along with custody changes to a blockchain visible to manufacturers, logistics providers, the partnering insurer, and receiving hospitals, thereby creating a single, tamper-proof source of truth on product conditions [41].

This real-time transparency shifts accountability at every hand-off as any temperature excursion is immediately recorded, letting recipients reject compromised shipments, pinpointing exactly when and where the breach occurred, and preventing actors from falsifying logs. In turn, claim processing accelerates because the insurer already possesses verified evidence, while consistently compliant routes can earn premium reductions, incentivizing careful handling throughout the chain.

Qualitative reports from the pilot noted fewer delays and out-of-tolerance incidents, as carriers, who are aware their performance was indelibly logged, became more diligent [71]. Yet, sensitive commercial details like drug formulas remained private. Scaling opportunities include onboarding additional routes, logistics partners, and regulators, dovetailing with statutory mandates such as the U.S. FDA's

DSCSA for serialization and anti-counterfeiting, and expanding sensor inputs to monitor broader quality parameters.

Future smart contract layers could automate quarantines or recalls the instant a breach is logged, instantly alerting all stakeholders and pulling affected batches from inventory, thus elevating patient safety while further streamlining cost, compliance, and performance in pharmaceutical supply chains.

3. *Blockchain for global shipping and logistics (maritime transport sector).* TradeLens, the blockchain-powered platform co-developed by Maersk and IBM in 2018, set out to digitize and connect the notoriously opaque global container-shipping ecosystem by recording real-time vessel departures and arrivals, port gate moves, customs releases, and electronic bills of lading on permissioned Hyperledger Fabric ledger that granted shippers, carriers, ports, customs, and logistics providers role-based access [35].

By 2020, five of the world's six largest ocean carriers, hundreds of ports, and more than 300 organizations were contributing data, ultimately logging more than a billion shipping events. This slashed document-handling times from days to mere minutes, gave customs instant manifest visibility, and enabled stakeholders to adjust downstream plans the moment a delay surfaced [13]. The shared, immutable record curbed fraud and disputes over cargo status such in the case of temperature-controlled loads, while trimming reliance on intermediaries, and the consensus mechanism meant no single actor, not even Maersk, could tamper with records. This strengthened trust across fiercely competitive players [33].

Yet governance concerns and uneven adoption as some carriers and governments were balking at perceived Maersk control hindered the network effect. In late 2022, Maersk and IBM shuttered TradeLens, citing insufficient industry-wide commitment to sustain the platform [47].

The experience spotlights future opportunities such as structuring more neutral or consortium-led governance to encourage participation, layering trade-finance services such as automated invoice factoring onto the rich data stream, extending coverage seamlessly across multimodal legs (truck, rail, and air), and boosting throughput and scalability to handle truly global volumes. All of these aspects could revive the promise of blockchain-enabled transparency and efficiency in international trade.

4. *Blockchain for managing the path of diamonds and wine (luxury goods sector).* Everledger's blockchain platform illustrates how immutable digital ledgers, enriched with high-resolution imagery, laser inscriptions, and later IoT devices, can safeguard provenance and authenticity for high-value, fraud-prone assets, beginning with diamonds and expanding into fine wines and luxury goods [66]. By creating a unique digital identity for each item and capturing every custody change, from mine to cutter to retailer for diamonds, or from cellar to auction for wine, Everledger transforms the paper-based Kimberley Process and traditional certificates into tamper-proof, instantly searchable records that consumers, insurers, and regulators can trust [28].

In diamonds, this transparency helps keep conflict stones out of legitimate markets. In other words, any stone lacking a matching blockchain record or showing inconsistent attributes is flagged, protecting ethical businesses and simplifying law-enforcement efforts [66]. In wine, a tamper-evident RFID cork tag links more than 90 data points (winery, vintage, photos, ownership history) to an allowed blockchain so buyers can verify a bottle's journey and authenticity with a quick scan [53]. This is especially vital in a secondary market where up to one in five bottles were once estimated to be counterfeit.

The shared ledger also streamlines certification, shifting verification from manual expert appraisal to rapid scan-and-verify processes, lowering insurance costs, preserving resale value, and automating performance tracking (Schmidt et al., 2021). Recognizing that provenance alone is insufficient for quality-sensitive goods, Everledger has begun integrating sensors to log temperature deviations or cap tampering, thereby monitoring storage and opening events in real-time and enhancing performance management [69].

Beyond gems and wine, the model's adaptability invites expansion into art, fashion, electronics, and sustainably sourced materials, though widespread impact hinges on persuading all major industry players to join, balancing transparency with data privacy through permissioned access and owner-controlled disclosure.

To synthesize the insights from the four cases, we compare them along several key dimensions: the specific blockchain application/purpose, the strengths or benefits realized, and the opportunities for growth or improvement identified. Table 10.1 presents a comparative framework summarizing these aspects for each case.

Across these diverse cases, common strengths emerge. Blockchain technology consistently delivered greater visibility and traceability, leading to faster response times when issues arose (as seen in Walmart's food recalls and TradeLens's shipping updates). It also built trust by providing data immutability, which in turn reduced the need for middlemen and manual verifications (e.g., Everledger bypassing traditional certifiers for diamonds, TradeLens reducing document couriers). These improvements are often translated into cost savings or performance gains, such as lower recall costs, less counterfeit loss, or streamlined administrative work. In terms of opportunities, a unifying theme is scaling and integration, meaning scaling the networks to include more participants and achieve critical mass, and integrating blockchain data with other systems and technologies such as IoT, artificial intelligence (AI) analytics, and ERP systems. The cases also show that blockchain projects must evolve toward standardization and collaboration. For instance, the need for neutral governance in TradeLens, or industry standards for tags in luxury goods tracking, is pinpointed as crucial for realizing this technology's full potential.

Table 10.1 Comparative framework of blockchain applications in supply chain cases (strengths, opportunities, and specific uses)

Case/company (sector)	Blockchain application	Strengths (observed benefits)	Opportunities for growth
Food and retail	Product traceability for food safety (IBM Food Trust blockchain for produce, pork)	– Dramatically faster traceability (weeks into seconds) enabling targeted recalls – Improved transparency and trust among supply chain partners and consumers (access to origin and handling data) – Reduced waste and recall costs by pinpointing affected batches instead of broad recalls	– Expand to more product categories and suppliers (industry-wide adoption for all food items) – Integrate IoT for automatic data capture (e.g., farm sensors uploading to blockchain) to enhance data richness and accuracy – Interoperate with regulators' systems for automated compliance reporting (food safety authorities accessing data directly)
Pharmaceutical sector and healthcare	Cold-chain monitoring for temperature-sensitive drugs (blockchain logging of temperature & custody data)	– Ensured drugs remained within safe conditions and immutable records of temperature at each handover – Greater accountability as every party's performance (maintaining the cold chain) is transparent, reducing disputes – Faster insurance claims and audits: blockchain data provided proof for compliance, speeding up claim settlements	– Scale up to more routes, products, and participants (global vaccine supply chains, etc.) – Add more sensor data (humidity, shock indicators) to monitor additional quality parameters on the blockchain – Employ smart contracts to automatically flag and quarantine out-of-compliance shipments (real-time quality control triggers)

(continued)

Table 10.1 (continued)

Case/company (sector)	Blockchain application	Strengths (observed benefits)	Opportunities for growth
Maritime transport	Global shipping platform for container tracking and document exchange via blockchain	– Improved efficiency in shipment visibility and document workflows (reducing processing times from days to hours) – High-industry adoption; 5 of 6 top ocean carriers joined, covering more than 50% of container traffic, creating significant network data effects – Enhanced data trust and security as a single shared ledger reduced reliance on intermediaries and chances for fraud in documentation	– Increase ecosystem adoption: bring on remaining carriers, forwarders, and governments (greater network completeness) – Adjust governance to be more neutral (ease competitors' concerns) and involve independent consortium governance to sustain participation – Extend to full intermodal supply chain (integrating air, rail, and truck blockchain data) for end-to-end logistics visibility
Luxury goods	Provenance tracking for high-value goods (diamonds, fine wine) using blockchain identities and tags	– Strong anti-counterfeiting: significantly reduced fraud by providing immutable provenance (e.g., diamond history, wine origin) – Increased customer trust and brand protection as buyers can verify authenticity and ethical sourcing (important for luxury markets) – Streamlined certification processes because automated verification of provenance replaces manual expert inspections in many cases	– Incorporate condition monitoring (IoT sensors for storage conditions) to complement provenance with quality data – Expand to other luxury or critical items (e.g., art, fashion, electronics for warranty) to broaden the platform's use – Achieve wider industry adoption by onboarding more producers, certification bodies, and marketplaces; possibly develop industry standards for tagging and blockchain records

Source Authors' analysis

10.6 Future of Blockchain in Supply Chain Tracking and Management

Looking ahead, the future of blockchain in supply chain management appears promising, with several key trends and developments likely to shape its evolution. One prominent trend is the convergence of blockchain with other advanced technologies to create smarter and more autonomous supply chains [14]. For example, the integration of IoT sensors with blockchain networks is expected to deepen. IoT devices can serve as automatic data providers to blockchains, feeding real-time information on location, temperature, humidity, etc., for assets in transit. This synergy can further reduce human intervention and error in data capture, thus enhancing reliability. As one analysis noted, combining blockchain with IoT can enhance trust and transparency while accelerating processes and reducing human intervention in supply chain operations [37]. This would effectively create self-tracking shipments and enable what some call "Industry 4.0" or the autonomous supply chain, where many transactions and decisions are executed automatically based on sensor data and smart contract logic [31].

While these integrations are still in the early stages, another technology convergence is with AI and data analytics [21]. With blockchains amassing large volumes of supply chain data, which is largely trustworthy and structured, AI can be applied to identify patterns, predict disruptions, and optimize operations. For example, machine learning algorithms could analyze blockchain data to predict demand surges or potential supplier delays, and smart contracts could automatically adjust orders or logistics plans accordingly. Some researchers have proposed frameworks where blockchain provides the data infrastructure and AI provides the decision intelligence, together enabling predictive and adaptive supply chain management [68]. In the case of the Maersk TradeLens project, a concept emerged to integrate an AI-based risk system that, using blockchain data, could dynamically reroute shipments if a port was suddenly closed as a form of automated resilience.

Standardization and interoperability will likely be a focus in the coming years. Currently, multiple blockchain platforms and consortia exist [62]. The future may hold cross-chain solutions or unified standards that allow data to be shared across different blockchains. For example, an international standards body might define data schema for supply chain events on the blockchain, or new protocols might emerge to facilitate interoperability. Connected to this, projects in the blockchain space often discuss "interoperability layers" or the use of intermediary services to bridge blockchains [9]. Interoperability ensures that blockchain's benefits are not confined to silos but rather span the end-to-end chain even when multiple platforms are involved.

Another area to watch is regulatory and legal frameworks catching up to blockchain. As blockchain tracking becomes more prevalent, regulators may start to accept or even mandate blockchain records for compliance [74]. For example, food safety regulators might allow companies to simply grant access to their blockchain traceability data instead of filing lengthy compliance reports. This would reduce administrative overhead and improve regulators' ability to pinpoint issues. Similarly, customs and trade authorities might integrate with shipping blockchains to automate trade clearance. There are already discussions in international bodies like WTO and WCO about embracing electronic documents, and blockchain could accelerate that shift by providing authenticity [5]. Governments might also encourage blockchain for public-good objectives, for instance, to ensure conflict-free minerals or to improve drug supply transparency in public health crises referring to some pilot projects that tracked COVID-19 vaccine distribution with blockchain for transparency [52].

From a performance measurement perspective, blockchain is likely to enable more granular and real-time KPIs for supply chains. We may see the emergence of new performance indicators that were not practical before. For example, "traceability time" as demonstrated by retailers could become a standard KPI in food industry scorecards, or the percentage of products with full provenance data available might be a metric consumers see [49]. Organizational performance measurements might include metrics like the supply chain trust index [16], which this time would be measured by factoring in blockchain-tracked sources.

Challenges remain part of the future scenario as well. Energy consumption concerns of earlier blockchains are mitigated by moving away from proof-of-work to greener consensus mechanisms, aligning blockchain with corporate sustainability goals [7]. Data privacy will continue to be important as techniques like zero-knowledge proofs may be employed so that participants can validate certain data without revealing sensitive details, which could allow even competitors to share a blockchain, without exposing their trade secrets [76].

The future will likely validate the multi-industry relevance of blockchain with more examples of adoption in sectors not deeply covered yet. We might see blockchain tracking in fashion (for verifying sustainable textiles, as already piloted by some brands), in energy (tracking renewable energy credits or battery supply chains), in electronics (the provenance of components to fight counterfeit electronics, which is a big issue for aerospace and defense supply chains). The common thread is reducing intermediary reliance, improving transparency, and enhancing performance measurement. Blockchain is poised to be a backbone technology that underpins trust in complex networks, much like the internet underpins communication. As costs of implementation decrease and success stories accumulate, more companies will incorporate blockchain into their digital supply chain strategies, not as a buzzword, but as a practical infrastructure for achieving resilient and transparent operations.

10.7 Conclusion

Blockchain technology is catalyzing a paradigm shift in supply chain tracking and performance management. As this chapter has detailed, blockchain's core properties of decentralization, immutability, and transparency directly address many of the longstanding pain points in complex supply networks. By providing a shared, tamper-proof ledger of transactions and events, blockchain reduces the need for intermediaries that were traditionally required to broker trust between parties [38]. Subsequently, this reduced intermediary reliance streamlines processes and minimizes opportunities for errors and fraud, as every stakeholder can verify data on the ledger for themselves [78]. The result is a leaner supply chain where information flows swiftly and securely to those who need it.

A recurring theme in our analysis is the dramatic improvement in transparency and traceability that blockchain enables. Whether tracking mangos, medicine, shipping containers, diamonds, or metals, the cases demonstrated that blockchain can offer end-to-end visibility at a granularity and speed previously unattainable. This transparency builds trust as consumers trust that the product they buy is authentic and safe, businesses trust their suppliers' data, and regulators trust the integrity of compliance information. In today's global market, such trust is a key component of performance. For instance, supply chain transparency is increasingly linked to brand reputation and customer satisfaction and blockchain provides a tool to bolster both by openly showcasing provenance and handling data. Furthermore, transparency enhances organizational performance measurement as firms can now measure things like traceability speed, chain-of-custody integrity, and supplier compliance in quantitative ways, using the rich data from blockchains as a source. These metrics were either impossible or very clumsy to capture before, but now they can be monitored continuously, enabling a more proactive and data-driven approach to managing supply chain performance.

The multi-industry cases demonstrate that the benefits of blockchain in supply chains are not confined to a single sector. From agriculture to pharma to logistics to luxury goods, we saw a common pattern where blockchain implementations led to more timely and accurate information, which in turn led to better decisions and outcomes. A supermarket can respond to food contamination in seconds rather than days, a hospital can trust the integrity of a vaccine shipment, a shipping company can automate its documentation, and a jeweler can guarantee a diamond's origin. In essence, blockchain is becoming a performance management tool as much as an IT tool, enabling what we might call transparent performance management.

That said, this chapter also illuminated the challenges and lessons learned. Not all blockchain projects have fully succeeded, teaching us that technology alone is not a silver bullet. Effective implementation requires careful attention to governance, stakeholder incentives, data privacy, and integration with existing processes. It requires industries to collaborate on standards and to sometimes cede a degree of control in exchange for mutual benefits. The path to widespread adoption will depend on addressing these challenges.

In conclusion, blockchain's role in supply chain management is poised to grow from niche pilot projects to industry-wide infrastructures. By removing friction and building trust, blockchain empowers supply chains to be more responsive, accountable, and resilient. It enhances visibility which improves coordination, it provides reliable data which improves decision-making, and it automates trust which frees organizations from cumbersome manual oversight. In a world where consumers and governments alike are demanding transparency and responsibility, blockchain is emerging as a key enabler of supply chains that are faster, cheaper, and better in the broadest sense. The academic and practical explorations synthesized in this chapter affirm that supply chain tracking and blockchain as a performance management tool is not a futuristic concept but an evolving reality that promises to redefine global business.

References

1. Abdallah, A.B., Rawadiah, O.M., Al-Byati, W., Alhyari, S.: Supply chain integration and export performance: the mediating role of supply chain performance. Int. J. Product. Perform. Manag. **70**(7), 1907–1929 (2021)
2. Agrawal, T.K., Angelis, J., Khilji, W.A., Kalaiarasan, R., Wiktorsson, M.: Demonstration of a blockchain-based framework using smart contracts for supply chain collaboration. Int. J. Prod. Res. **61**(5), 1497–1516 (2023)
3. Akram, W., Joshi, R., Haider, T., Sharma, P., Jain, V., Garud, N., Singh, N.: Blockchain technology: A potential tool for the management of pharma supply chain. Res. Social Adm. Pharm. **20**(6), 156–164 (2024)
4. Alkhudary, R., Féniès, P.: Blockchain and trust in supply chain management: a conceptual framework. IFAC-PapersOnLine **55**(10), 2402–2406 (2022)
5. Allen, D.W., Berg, C., Davidson, S., Novak, M., Potts, J.: International policy coordination for blockchain supply chains. Asia Pac. Policy Stud. **6**(3), 367–380 (2019)
6. Allenbrand, C.: Smart contract-enabled consortium blockchains for the control of supply chain information distortion. Blockchain: Res. Appl. **4**(3), 100134 (2023)
7. Almutairi, K., Hosseini Dehshiri, S.J., Hosseini Dehshiri, S.S., Hoa, A.X., Arockia Dhanraj, J., Mostafaeipour, A., Techato, K., et al.: Blockchain technology application challenges in renewable energy supply chain management. Environ. Sci. Pollut. Res. **30**(28), 72041–72058 (2023)
8. Al-Rakhami, M.S., Al-Mashari, M.: A blockchain-based trust model for the internet of things supply chain management. Sensors **21**(5), 1759 (2021)
9. Al-Rakhami, M., Al-Mashari, M.: Interoperability approaches of blockchain technology for supply chain systems. Bus. Process. Manag. J. **28**(5/6), 1251–1276 (2022)
10. Aslam, J., Saleem, A., Kim, Y.B.: Blockchain-enabled supply chain management: integrated impact on firm performance and robustness capabilities. Bus. Process. Manag. J. **29**(6), 1680–1705 (2023)
11. Azevedo, P., Gomes, J., Romão, M.: Supply chain traceability using blockchain. Oper. Manag. Res. **16**(3), 1359–1381 (2023)
12. Belchior, R., Vasconcelos, A., Guerreiro, S., Correia, M.: A survey on blockchain interoperability: past, present, and future trends. ACM Comput. Surv. (CSUR) **54**(8), 1–41 (2021)
13. Bjørn-Andersen, N., Kouwenhoven, N.: Boosting the effectiveness of containerised supply chains: a case study of TradeLens. In: Maritime Informatics: Additional Perspectives and Applications, vol. 95 (2021)
14. Borah, M.D., Naik, V.B., Patgiri, R., Bhargav, A., Phukan, B., Basani, S.G.: Supply chain management in agriculture using blockchain and IoT. In: Advanced Applications of Blockchain Technology, pp. 227–242. Springer Singapore, Singapore (2019)
15. Bovenzi, G., Aceto, G., Persico, V., Pescapé, A.: Blockchain Performance in Industry 4.0: drivers, use cases, and future directions. J. Ind. Inform. Integr. **36**, 100513 (2023)
16. Capaldo, A., Giannoccaro, I.: Interdependence and network-level trust in supply chain networks: a computational study. Ind. Mark. Manage. **44**, 180–195 (2015)
17. Casino, F., Dasaklis, T.K., Patsakis, C.: A systematic literature review of blockchain-based applications: current status, classification and open issues. Telemat. Inform. **36**, 55–81 (2019)

18. Castka, P., Searcy, C., Mohr, J.: Technology-enhanced auditing: Improving veracity and timeliness in social and environmental audits of supply chains. J. Clean. Prod. **258**, 120773 (2020)
19. Centobelli, P., Cerchione, R., Del Vecchio, P., Oropallo, E., Secundo, G.: Blockchain technology for bridging trust, traceability and transparency in circular supply chain. Inform. Manage. **59**(7), 103508 (2022)
20. Chang, S.E., Chen, Y.C., Lu, M.F.: Supply chain re-engineering using blockchain technology: a case of smart contract based tracking process. Technol. Forecast. Soc. Chang. **144**, 1–11 (2019)
21. Charles, V., Emrouznejad, A., Gherman, T.: A critical analysis of the integration of blockchain and artificial intelligence for supply chain. Ann. Oper. Res. **327**(1), 7–47 (2023)
22. Cole, R., Stevenson, M., Aitken, J.: Blockchain technology: implications for operations and supply chain management. Supply Chain Manage. Int. J. **24**(4), 469–483 (2019)
23. Daghighi, A., Shoushtari, F.: Toward sustainability of supply chain by applying blockchain technology. Int. J. Indus. Eng. Oper. Res. **5**(2), 60–72 (2023)
24. De Giovanni, P.: Blockchain and smart contracts in supply chain management: a game theoretic model. Int. J. Prod. Econ. **228**, 107855 (2020)
25. Farshidi, S., Jansen, S., España, S., Verkleij, J.: Decision support for blockchain platform selection: three industry case studies. IEEE Trans. Eng. Manage. **67**(4), 1109–1128 (2020)
26. Ghadge, A., Bourlakis, M., Kamble, S., Seuring, S.: Blockchain implementation in pharmaceutical supply chains: a review and conceptual framework. Int. J. Prod. Res. **61**(19), 6633–6651 (2023)
27. Godavarthi, B., Dhar, M., Devi, S.A., Raju, S.S., Balaram, A., Srilakshmi, G.: Blockchain integration with the internet of things for the employee performance management. J. High Technol. Managem. Res. **34**(2), 100468 (2023)
28. Gordon, S.R., Gilleran, R., Shankaranarayanan, G., Stoddard, D.B., Johnson, B.A.: Blockchain adoption in the supply chain—a game theoretic perspective: the case of the diamond industry. J. Inform. Technol. Case Appl. Res. **25**(3), 241–276 (2023)
29. Gurtu, A., Johny, J.: Potential of blockchain technology in supply chain management: a literature review. Int. J. Phys. Distrib. Logist. Manag. **49**(9), 881–900 (2019)
30. Helo, P., Shamsuzzoha, A.H.M.: Real-time supply chain—a blockchain architecture for project deliveries. Robot. Comput.-Integr. Manuf. **63**, 101909 (2020)
31. Hofmann, E., Sternberg, H., Chen, H., Pflaum, A., Prockl, G.: Supply chain management and Industry 4.0: conducting research in the digital age. Int. J. Phys. Distrib. Logist. Manage. **49**(10), 945–955 (2019)
32. Jackson, A., Spiegler, V.L., Kotiadis, K.: Exploring the potential of blockchain-enabled lean automation in supply chain management: a systematic literature review, classification taxonomy, and future research agenda. Prod. Plan. Control **35**(9), 866–885 (2024)
33. Jensen, T., Hedman, J., Henningsson, S.: How TradeLens delivers business value with blockchain technology. MIS Quarter. Execut. **18**(4) (2019)
34. Jeyanthi, P.M.: Theories of cryptocurrency, blockchain and distributed systems and environmental implications. In: Cryptocurrencies and Blockchain Technology Applications, pp. 215–238 (2020)
35. Jovanovic, M., Kostić, N., Sebastian, I.M., Sedej, T.: Managing a blockchain-based platform ecosystem for industry-wide adoption: the case of TradeLens. Technol. Forecast. Soc. Chang. **184**, 121981 (2022)
36. Katsikouli, P., Wilde, A.S., Dragoni, N., Høgh-Jensen, H.: On the benefits and challenges of blockchains for managing food supply chains. J. Sci. Food Agric. **101**(6), 2175–2181 (2021)
37. Kaur, A., Singh, G., Kukreja, V., Sharma, S., Singh, S., Yoon, B.: Adaptation of IoT with blockchain in Food Supply Chain Management: an analysis-based review in development, benefits and potential applications. Sensors **22**(21), 8174 (2022)
38. Khan, M.Z., Kumar, A., Sahu, A.K.: Blockchain applications in supply chain management: a systematic review of reviews. Glob. Knowl. Mem. Commun. **74**(3/4), 1191–1208 (2025)

39. Khan, S.N., Loukil, F., Ghedira-Guegan, C., Benkhelifa, E., Bani-Hani, A.: Blockchain smart contracts: applications, challenges, and future trends. Peer-to-peer Netw. Appl. **14**, 2901–2925 (2021)
40. Kshetri, N.: Blockchain's roles in meeting key supply chain management objectives. Int. J. Inf. Manage. **39**, 80–89 (2018)
41. Kshetri, N., Voas, J.: Supply chain trust. IT Professional **21**(2), 6–10 (2019)
42. Kumar, A., Liu, R., Shan, Z.: Is blockchain a silver bullet for supply chain management? Technical challenges and research opportunities. Decis. Sci. **51**(1), 8–37 (2020)
43. Latif, R.M.A., Farhan, M., Rizwan, O., Hussain, M., Jabbar, S., Khalid, S.: Retail level Blockchain transformation for product supply chain using truffle development platform. Clust. Comput. **24**, 1–16 (2021)
44. Lee, C.H., Yang, H.C., Wei, Y.C., Hsu, W.K.: Enabling blockchain based SCM systems with a real time event monitoring function for preemptive risk management. Appl. Sci. **11**(11), 4811 (2021)
45. Li, J., Maiti, A., Springer, M., Gray, T.: Blockchain for supply chain quality management: challenges and opportunities in context of open manufacturing and industrial internet of things. Int. J. Comput. Integr. Manuf. **33**(12), 1321–1355 (2020)
46. Longo, F., Nicoletti, L., Padovano, A., d'Atri, G., Forte, M.: Blockchain-enabled supply chain: an experimental study. Comput. Ind. Eng. **136**, 57–69 (2019)
47. Louw-Reimer, J., Nielsen, J.L.M., Bjørn-Andersen, N., Kouwenhoven, N.: Boosting the effectiveness of containerised supply chains: a case study of tradelens. In: Maritime informatics: additional perspectives and applications, pp. 95–115. Springer International Publishing, Cham (2021)
48. Mangla, S.K., Kazançoğlu, Y., Yıldızbaşı, A., Öztürk, C., Çalık, A.: A conceptual framework for blockchain-based sustainable supply chain and evaluating implementation barriers: a case of the tea supply chain. Bus. Strateg. Environ. **31**(8), 3693–3716 (2022)
49. Marinagi, C., Reklitis, P., Trivellas, P., Sakas, D.: The impact of industry 4.0 technologies on key performance indicators for a resilient supply chain 4.0. Sustainability **15**(6), 5185 (2023)
50. Markus, S., Buijs, P.: Beyond the hype: how blockchain affects supply chain performance. Supply Chain Manag. Int. J. **27**(7), 177–193 (2022)
51. Nakasumi, M.: Information sharing for supply chain management based on block chain technology. In: 2017 IEEE 19th Conference on Business Informatics (CBI), Vol. 1, pp. 140–149. IEEE (2017, July)
52. Omar, I.A., Debe, M., Jayaraman, R., Salah, K., Omar, M., Arshad, J.: Blockchain-based supply chain traceability for COVID-19 personal protective equipment. Comput. Ind. Eng. **167**, 107995 (2022)
53. Parry, G.M., Revolidis, I., Ellul, J., Pace, G.: Bottling up trust: a review of blockchain adoption in wine supply chain traceability. IEEE Access (2024)
54. Queiroz, M.M., Telles, R., Bonilla, S.H.: Blockchain and supply chain management integration: a systematic review of the literature. Supply Chain Manage. Int. J. **25**(2), 241–254 (2020)
55. Reddy, K.R.K., Gunasekaran, A., Kalpana, P., Sreedharan, V.R., Kumar, S.A.: Developing a blockchain framework for the automotive supply chain: a systematic review. Comput. Ind. Eng. **157**, 107334 (2021)
56. Rogerson, M., Parry, G.C.: Blockchain: case studies in food supply chain visibility. Supply Chain Manage. Int. J. **25**(5), 601–614 (2020)
57. Saberi, S., Kouhizadeh, M., Sarkis, J., Shen, L.: Blockchain technology and its relationships to sustainable supply chain management. Int. J. Prod. Res. **57**(7), 2117–2135 (2019)
58. Sahoo, S., Kumar, S., Sivarajah, U., Lim, W.M., Westland, J.C., Kumar, A.: Blockchain for sustainable supply chain management: trends and ways forward. Electron. Commer. Res. **24**(3), 1563–1618 (2024)
59. Schmidt, C.G., Klöckner, M., Wagner, S.M.: Blockchain for supply chain traceability: case examples for luxury goods. In: Digital Business Models in Industrial Ecosystems: Lessons Learned from Industry 4.0 Across Europe, pp. 187–197 (2021)

60. Sharif, M.M., Ghodoosi, F.: The ethics of blockchain in organizations. J. Bus. Ethics **178**(4), 1009–1025 (2022)
61. Sharma, M., Kumar, P.: Adoption of blockchain technology: a case study of Walmart. In: Blockchain Technology and Applications for Digital Marketing, pp. 210–225. IGI Global (2021)
62. Shoaib, M., Zhang, S., Ali, H.: A bibliometric study on blockchain-based supply chain: a theme analysis, adopted methodologies, and future research agenda. Environ. Sci. Pollut. Res. **30**(6), 14029–14049 (2023)
63. Singh, V., Sharma, S.K.: Application of blockchain technology in shaping the future of food industry based on transparency and consumer trust. J. Food Sci. Technol. **60**(4), 1237–1254 (2023)
64. Song, W., Zhu, M., Lu, D., Zhu, C., Zhao, J., Sun, Y., Zhu, H., et al.: Blockchain bottleneck analysis based on performance metrics causality. Electronics **13**(21), 4236 (2024)
65. Tan, B., Yan, J., Chen, S., Liu, X.: The impact of blockchain on food supply chain: The case of walmart. In: Smart Blockchain: First International Conference, SmartBlock 2018, Tokyo, Japan, December 10–12, 2018, Proceedings, vol. 1, pp. 167–177. Springer International Publishing (2018)
66. Thakker, U., Patel, R., Tanwar, S., Kumar, N., Song, H.: Blockchain for diamond industry: opportunities and challenges. IEEE Internet Things J. **8**(11), 8747–8773 (2020)
67. Tian, F.: An agri-food supply chain traceability system for China based on RFID & blockchain technology. In: 2016 13th International Conference on Service Systems and Service Management (ICSSSM), pp. 1–6. IEEE (2016, June)
68. Tsolakis, N., Schumacher, R., Dora, M., Kumar, M.: Artificial intelligence and blockchain implementation in supply chains: a pathway to sustainability and data monetisation? Ann. Oper. Res. **327**(1), 157–210 (2023)
69. Vazquez Melendez, E.I., Bergey, P., Smith, B.: Blockchain technology for supply chain provenance: increasing supply chain efficiency and consumer trust. Supply Chain Manage. Int. J. **29**(4), 706–730 (2024)
70. Vu, N., Ghadge, A., Bourlakis, M.: The impact of Blockchain adoption on supply chain performance: evidence from food industry. Int. J. Prod. Res. 1–26 (2024)
71. Wade, D., Kaniki, F.R., Ayosanmi, O.S., Faith, A.O., Olusegun, A., Felix, S.O., Musa, A. et al.: Blockchain and big data analytics in the optimization of Nigeria vaccine supply chain. GSJ **7**(11) (2019)
72. Wahab, S.N., Hamzah, M.I., Sohal, A.: Leveraging top management support for blockchain-driven innovations in logistics and supply chain. Asia-Pac. J. Bus. Admin. (2025)
73. Wamba, S.F., Queiroz, M.M.: Blockchain in the operations and supply chain management: Benefits, challenges and future research opportunities. Int. J. Inf. Manage. **52**, 102064 (2020)
74. Wang, B., Lin, Z., Wang, M., Wang, F., Xiangli, P., Li, Z.: Applying blockchain technology to ensure compliance with sustainability standards in the PPE multi-tier supply chain. Int. J. Prod. Res. **61**(14), 4934–4950 (2023)
75. Wieland, A.: Dancing the supply chain: Toward transformative supply chain management. J. Supply Chain Manag. **57**(1), 58–73 (2021)
76. Xu, P., Lee, J., Barth, J.R., Richey, R.G.: Blockchain as supply chain technology: considering transparency and security. Int. J. Phys. Distrib. Logist. Manag. **51**(3), 305–324 (2021)
77. Yang, J., Ma, X., Crespo, R.G., Martínez, O.S.: Blockchain for supply chain performance and logistics management. Appl. Stoch. Model. Bus. Ind. **37**(3), 429–441 (2021)
78. Yavaprabhas, K., Pournader, M., Seuring, S.: Blockchain as the "trust-building machine" for supply chain management. Ann. Oper. Res. **327**(1), 49–88 (2023)

Chapter 11
Blockchain-Enabled Cross-Asset Finance: Tokenization, AI-Driven Trading, and Quantum-Resilient Auditing

Siddharth Pareek

Abstract Modern finance faces three critical gaps: fragmented asset classes, unequal access to advanced analytics, and episodic, labour-intensive audits. In 2023, the global payments industry processed 3.4 trillion transactions valued at $1.8 quadrillion, generating $2.4 trillion in revenue—yet settlement often still takes two days and fees average 1–2% per transaction. Over 130 million retail investors now trade via mobile apps—up from 51.9 million in 2019—but just 15 per cent leverage AI-driven analytics; the algorithmic trading market remains an institutional stronghold at $21 billion in. Meanwhile, the global auditing services market reached $227 billion in 2023, with audits still conducted quarterly or annually, leaving extended windows of undetected risk. GADE 2.0: A cross-asset, blockchain-native exchange unifying tokenized equities, commodities, and cryptocurrencies (including over-collateralized synthetic assets) on a PoS/L2 chain that settles trades in fully reserved stablecoins within seconds—slashing fees by up to 90% and collapsing T + 2 settlement into sub-second finality. Democratized AI-Driven Hub: An end-to-end investment and trading platform ingesting terabytes of live on-chain and market data, deploying ensemble ML and reinforcement learning models to generate retail-grade predictive signals, automated smart-contract executions, and dynamic hedging—extending institutional-grade analytics to every investor worldwide. Quantum-Secure Continuous Audit: A next-gen audit framework embedding NIST-standard post-quantum cryptography into the ledger, augmented by AI-driven anomaly detection running continuously on transaction flows. This ensures millisecond-scale assurance mapped to PCAOB/IFRS frameworks, eliminating audit blind spots and future-proofing against quantum threats projected by 2030–2035.

Keywords Blockchain tokenization · Artificial intelligence (AI) in finance · Quantum-resilient cryptography · Distributed ledger technology (DLT) · Cross-asset trading platforms · Real-time continuous auditing · GADE 2.0

S. Pareek (✉)
ICICI AMC, Connaught Place, New Delhi, Delhi, India
e-mail: Siddharth_Pareek@icicipruamc.com

P. M. Jeyanthi (ed.), *Distributed Ledger Technology in Communication*, Signals and Communication Technology, https://doi.org/10.1007/978-3-032-07682-3_11

architecture · Democratized financial analytics · Post-quantum security solutions · Autonomous and AI-driven financial markets

11.1 Introduction

Financial markets today are vast but still rely on slow, siloed processes. Global financial markets now exceed **$1088 trillion** in notional value, a record highly driven by derivatives and cash equities. At the same time, the payment industry is processing an enormous volume of transactions, roughly **3.4 trillion payments** worth **$1.8 quadrillion** in value during 2023. However, many cross-border payments and trades still take hours or days to clear, and manual audits of these complex transactions occur only periodically. This lag creates large pools of illiquid funds and hidden risks. Institutions have access to advanced analytics and AI tools (for example, leading banks report ~ 20% productivity gains from AI in areas like coding and customer service, whereas retail investors and smaller firms often lack such capabilities. These dynamics—massive scale, growing data complexity, and unequal technology access—highlight critical gaps in today's financial system. The following sections outline these gaps in the current landscape before exploring how tokenization, AI-driven trading, and DLT-based audit methods can address them.

11.2 Current Landscape

The sheer size of markets and flows is striking, but frictions remain. For example:

Despite this scale, **settlement frictions** persist. Equity and bond trades commonly settle two business days after the trade date (T + 2), and many cross-border payments can take hours or days to clear. BIS data on SWIFT GPI show that the median end-to-end cross-border payment takes under 2 h, but the slowest routes exceed 2 days. Each extra intermediary or manual compliance check further delays final settlement and ties up liquidity.

There is also an **analytics and technology divide**. Large banks, hedge funds, and exchanges now deploy advanced AI and big-data systems, while most retail platforms and small firms remain on simpler tools. Surveys find that major financial institutions are heavily investing in AI (with many reporting ~ 20% efficiency gains in key processes, whereas individual investors and community firms do not have access to comparable algorithmic strategies. This creates a widening gap: sophisticated players can analyze and trade multiple asset classes rapidly, while others face "walls" of fragmented data and risk being unable to respond quickly.

Finally, **audit and compliance lags** are significant. Traditional audits are conducted after transactions and are often based on manual sample testing. As a result, errors or fraud may not be detected for months. One study notes that many organizations still have information "difficult to access or trust" and siloed processes that "restrict agility". In practical terms, yearly financial audits can lag far behind real time, leaving gaps in oversight.

These gaps—slow settlement and reconciliation, an analytics divide, and delayed audit feedback—all point to opportunities for distributed-ledger solutions. By enabling near-instant settlement (via tokenized assets), democratized data access, and continuous on-chain audit trails, blockchain-based finance can begin to bridge today's frictions and set the stage for the innovations discussed later in this chapter.

11.3 Review of Foundational Research

Recent work has shown that tokenizing real-world assets (RWAs) can significantly reshape financial markets. Tokenization frameworks are being explored globally through regulatory sandboxes and pilots. For example, initiatives like Singapore's Project Guardian and similar efforts in Europe and Hong Kong have spurred trials of tokenized bonds, equities and loans. Studies find that when assets are tokenized, ownership can be split into liquid digital shares—a change that can *democratize access* and improve liquidity. Riabokin and Kotukh argue that RWA tokenization represents a "paradigm shift", enabling broader investor participation and new financing models in recovering economies. Likewise, research on real estate tokenization emphasizes that fractional ownership makes high-value properties accessible to individual investors, potentially leading to more inclusive markets. These findings are supported by industry analyses: McKinsey & Company projects that tokenized financial assets could reach roughly $2 trillion in market value by 2030 (excluding pure crypto and stablecoins). In practice, major market infrastructures are already running pilots (for example, DTCC, Clearstream and Euroclear have announced blockchain test projects) and new platforms (e.g. JPMorgan's Onyx, Citigroup pilots) are being built to support on-chain asset trading.

Key Developments in Tokenization:

- **Regulatory sandboxes**: Governments are creating frameworks (e.g. Europe's MiCA, the UK's Digital Securities Sandbox, Singapore's Project Guardian) to clarify legal treatment of digital assets.
- **Infrastructure pilots**: Financial institutions are testing on-chain settlements and token issuance; projects include blockchain-based bond trials and tokenized fund platforms.
- **Research insights**: Academic surveys note that tokenization can *"democratize asset access, increasing the liquidity of previously illiquid assets"*. These works stress that clear legal definitions (distinguishing asset-backed tokens from service tokens) are essential to realize these benefits.

Research on AI and machine learning in trading is equally active. Quantitative finance is increasingly incorporating AI for prediction and strategy. Surveys report that deep learning models have boosted market forecasting accuracy and opened up new data sources (news, social media, alternative data). One recent review highlights that large language models (LLMs) can now process unstructured information to propose trading "alpha" factors, suggesting a shift from hand-crafted indicators to AI-driven signals. At the same time, reinforcement learning (RL) is being explored for automated trading strategies. While some experimental RL agents have shown promising performance on simplified tasks, regulators caution that full-scale deployment could introduce new risks. A legal analysis notes that widespread use of opaque deep learning or RL-based systems might create "monoculture" effects, raising systemic and market-manipulation concerns. In practice, most investment managers still rely on AI for analysis and support (rather than fully autonomous trading), but research indicates growing interest in creating AI agents that learn from market data over time.

Efforts to transform auditing with blockchain and AI have also been documented. Blockchain's immutable ledger and smart contracts can automate much of the audit trail. For example, PwC's prototype "networked audit system" connects corporate finance systems to a shared ledger. In one case study, syncing supplier, logistics, and customer data on-chain **reduced manual reconciliation time by 90%**, while an integrated AI model automatically flagged fictitious transactions totalling US$12 million. Similarly, Deloitte's pilot blockchain for trade finance (with JPMorgan) posted loan and repayment data on-chain and used smart contracts to verify compliance, slashing a letter-of-credit audit cycle from 14 to 2 days. These real-world examples illustrate how **smart contracts and ML** can turn periodic audits into continuous monitoring. Zhang et al. review this trend: they report that blockchain-enabled tools allow "complete data analysis" (instead of 5–10% sampling) and enable machine learning anomaly detection, while noting remaining technical and governance challenges. In summary, foundational work shows that combining DLT with AI can greatly speed up reconciliation and anomaly detection in financial records.

Finally, recent studies at the intersection of quantum computing and blockchain highlight the need for quantum-resistant solutions. Researchers agree that emerging quantum algorithms (e.g. Shor's and Grover's) can break current blockchain cryptography (like ECDSA signatures) and weaken proof-of-work security. For example, Parida et al. survey the field and stress that *"quantum algorithms pose a significant threat to both public-key cryptography and hash functions"*, implying a fundamental redesign of ledger protocols. In response, many proposals use post-quantum primitives: lattice-based and hash-based signature schemes (such as Falcon or SPHINCS+) have been evaluated for blockchain use. Some projects even prototype hybrid key-exchange protocols (combining classical and lattice schemes) to ease the transition. More recent work proposes **transition protocols** that softly upgrade existing blockchains to quantum-safe cryptography without disrupting ongoing operations. These studies provide a clear research foundation: a secure future DLT must integrate quantum-resistant algorithms and prepare migration strategies.

11.4 Why We Need a Finance Rebound

Despite these advances, the current financial infrastructure faces critical gaps that a "finance rebound" would address. One urgent need is **near-instant, low-cost settlement**. Today's markets often settle trades on T + 1 or T + 2 delays using intermediaries, which imposes counterparty risk and requires expensive collateral. Blockchain and DLT promise atomic settlement—payments and asset transfers that finalize in seconds. Instant settlement would dramatically reduce counterparty exposure and free up capital, aligning with regulators' calls for faster payment rails. By removing intermediaries, a DLT-based system could also cut transaction fees and speed up cross-border trades, strengthening market stability and liquidity.

Another challenge is **universal tokenization**. In theory any asset (stocks, bonds, commodities, real estate, even personal data) can be tokenized, but in practice several hurdles remain. *Liquidity* is a concern: many tokenized assets currently trade on nascent platforms with low volume. Without sufficient buyers and sellers, token prices can swing wildly or stalls. *Regulation* is also fragmented: jurisdictions have different rules on digital securities, often leaving businesses uncertain about cross-border compliance. Moreover, *usability* matters: for tokenized markets to take off, the ecosystem must be user-friendly. Wallets, exchanges, and smart contracts need standardized interfaces so that regulators, custodians, and end-users can confidently verify what a token represents and how to redeem it. Research on tokenization consistently points out these gaps—for instance, it lists **regulatory uncertainty, platform interoperability, privacy and legal enforceability** among the main obstacles. Overcoming them will require coordinated standards and tools to bridge on-chain assets with off-chain laws and data.

A third pillar is **democratizing AI-driven analytics**. Financial AI today largely benefits institutional investors; sophisticated models and data feeds are out of reach for most individuals. Yet over one hundred million retail investors worldwide are now active in markets (for example, stock-trading apps have seen explosive global growth). As the World Economic Forum reports, retail participation is rising thanks to "AI-driven platforms" and expanded access to financial products. To fully empower these individuals, we need to put advanced analytics in their hands. This means delivering AI tools (predictive signals, risk profiling, portfolio recommendations) through intuitive apps. Retail users should be able to tap the same real-time data processing and ML insights that firms use, levelling the playing field. In short, AI must become a public utility for finance, not just a proprietary advantage.

Finally, the mode of assuring financial integrity must change. Traditional audits are periodic and retrospective (e.g. annual or quarterly). This leaves long windows where errors or fraud can go unnoticed. A "continuous assurance" model—underpinned by DLT—could replace this cycle. On-chain records provide a permanent, tamper-evident trail of every transaction. When combined with AI anomaly detection, auditors could monitor in real time. For example, Deloitte's evolving practice now triggers *on-the-spot* reviews of any unusually large cash flow. In principle, an entire ledger could be analyzed continuously: Ernst & Young reported that its blockchain tools can raise risk coverage from 78% of transactions to nearly 100%. Continuous assurance would slash the delay between an event and its audit, reducing losses and boosting confidence. In summary, moving to blockchain-based, AI-enhanced audit mechanisms addresses the glaring inefficiency of waiting months to correct books.

11.5 Towards a New Platform: GADE 2.0 and Beyond

To address these needs, we envision a unified platform that integrates tokenization, AI, and quantum-safe security. **GADE 2.0** is proposed as a cross-asset, blockchain-native exchange. All asset classes (equities, fixed income, commodities, derivatives, etc.) would be tokenized and trade on a common ledger. Using a proof-of-stake (PoS) consensus and layer-2 scaling, GADE 2.0 could offer high throughput with low fees. Trades would settle instantly on-chain, eliminating counterparty risk and settlement lag. Crucially, GADE 2.0 would interoperate with existing finance: bridges to CBDC and fiat rails enable seamless conversion, and standardized token interfaces ensure regulatory compliance. This comprehensive DLT exchange would realize the promise of tokenization at scale, anchoring a resilient ecosystem.

Alongside GADE, we propose an **AI-Driven Investment Hub**. This subsystem would ingest live market data (prices, news, alternative feeds) in real time and feed it to AI/ML models. The models generate trading signals and risk analytics continuously. Smart-contract-based "trading bots" could then execute strategies automatically when pre-set conditions are met. Retail investors would access these features through intuitive dashboards and mobile apps. In effect, the same algorithms that professional desks use become transparent and configurable: for example, a user could allocate funds according to AI-derived momentum signals or portfolio optimization. By providing on-demand, automated analytics and execution, the AI Hub democratizes investment insights for millions of users.

Finally, a **quantum-secure audit layer** would lock in integrity and compliance. All on-chain transactions and smart contracts would use post-quantum cryptography (e.g. lattice-based signatures and hash-based proofs) to secure them against future quantum attacks. Audit logs would be duplicated on a parallel DLT network with quantum-safe encryption. An AI-driven monitoring engine would continuously scan the blockchain for anomalies or policy violations, much like modern intrusion detection. For privacy, zero-knowledge proofs would allow entities to prove compliance (for example, asset reserve levels or identity checks) without revealing underlying data. In combination, these measures provide "always-on" audit assurance that is future-proof against quantum threats.

In summary, the proposed architecture—GADE 2.0 with its AI and quantum-safe components—directly addresses the identified needs. It delivers atomic settlement and broad tokenization; it embeds AI analytics at the core; and it upgrades audit processes to continuous, secure monitoring. By bridging cutting-edge research with practical design, this solution framework could catalyze the finance rebound.

11.6 GADE 2.0: Blockchain-Enabled Cross-Asset Exchange

GADE 2.0 is envisioned as a high-throughput Layer-2/PoS-based platform that unifies multiple liquidity models and asset classes on one chain. This means a **hybrid liquidity model** combining on-chain Automated Market Makers (AMMs) with off-chain Central Limit Order Books (CLOBs). For example, platforms like Vertex illustrate this approach: their Arbitrum-based exchange integrates an on-chain AMM with a sequenced off-chain order book, allowing trades to "always [be] executed at the best available price" by tapping both pools and order-book liquidity. By meshing AMM pools and order-book orders in real time, GADE 2.0 can ensure deep liquidity even in low-volatility markets and dynamically route trades to the most efficient source. The underlying chain is secured by a proof-of-stake consensus (or similar L2 rollup) to maximize throughput and minimize fees, drawing on lessons from Ethereum's shift to PoS and the scalability of rollups (e.g. Arbitrum, Polygon) as its backbone.

GADE 2.0 targets **tokenized trading across asset classes**. It supports on-chain representations of equities, commodities, forex, cryptocurrencies, and synthetic assets, all tradable in one platform. In effect, traditional stocks, bonds, or commodities are "wrapped" into tokens or synthetics so they can be exchanged peer-to-peer on-chain. Regulators and advisors have noted that tokenizing real-world assets (like stocks or bonds) can boost "efficiency, transparency, liquidity, [and] accessibility", enabling things such as atomic cross-asset settlement. In parallel, established crypto protocols already demonstrate synthetics: Synthetix and mirror-type platforms let users mint tokenized derivatives ("Synths") for indices, fiat currencies, commodities, and other assets. By integrating these, GADE 2.0 could let a user trade, say, a tokenized S&P share against a gold token against a crypto like bitcoin, all in one smart order. External price oracles (e.g. Pyth Network) ingest real-world market data for equities, FX, and commodities, feeding GADE's on-chain oracles so that all asset prices stay up-to-date. This cross-asset functionality means GADE 2.0 serves as a one-stop multi-asset exchange, unifying markets that today are siloed by regulators and rails.

Settlement on GADE 2.0 leverages **stablecoin rails** to achieve near-instant finality and rock-bottom fees. By settling trades in on-chain dollar tokens (USDC, USDT, etc.) or similarly pegged currency tokens, the system bypasses slow legacy rails. As one industry analysis notes, blockchain stablecoins can settle payments "near instant globally" and operate 24/7. Empirically, on high-speed L1s like Solana, stablecoin transfers see "sub-second settlement times and minimal fees"—a performance game-changer compared to $T + 2 - T + 3$ stock trades or multi-day wire transfers. In practice GADE could take advantage of this: for example, instead of paying 0.3–0.5% in trading commissions and overnight settlement delays, trades on GADE might cost only a few basis points and settle in under a second, representing roughly a 90% fee reduction. (By contrast, traditional credit-card and bank fees often run into single-digit percentages.) In short, pegged tokens provide a digital rail that auto-settles across assets almost instantaneously. Finally, GADE 2.0 would embed *on-chain compliance* modules to pre-verify every participant under KYC/AML rules. For example, smart contracts could be coded to check proof of KYC or accredited-investor status (via on-chain attestations or trusted identity oracles) before allowing trading or minting of certain tokens. Emerging protocols like PureFi demonstrate how DeFi apps can require regulatory checks on-chain, suggesting GADE 2.0 could similarly enforce compliance conditions at the protocol level.

11.7 Democratized AI-Driven Investment and Trading Hub

In parallel, GADE 2.0 includes a retail-friendly **AI/algorithmic trading layer**. It continuously ingests real-time market data from both on-chain sources (DEX trades, chain metrics) and off-chain feeds (exchanges, market data APIs). For example, networks like the Pyth oracle already "deliver real-time market data for crypto, equities, FX, and commodities to 100 + blockchains". GADE's AI hub uses this constant data stream (along with economic news feeds, sentiment indices, etc.) to generate predictive signals via ensemble machine learning models and reinforcement learning. Such approaches are increasingly employed in quantitative trading: modern hedge funds use complex AI/ML stacks (including deep neural networks and ensemble methods) to model price patterns and portfolio behaviour. A GADE hub might run hundreds of models in parallel (e.g. XGBoost/random forest ensembles, recurrent networks, RL agents) to forecast short-term trends or volatility. These models produce buy/sell/hedge signals across assets, which are then accessible to users or algorithms via the platform.

Crucially, GADE 2.0 automates execution via **smart contracts**. Users can link signals to on-chain bots that place or manage orders without manual intervention. For instance, a smart contract could implement stop-loss and take-profit orders: if Token X drops below a set price, the contract automatically sells the user's position to limit losses. Another example is automated rebalancing: a portfolio contract could enforce a target asset allocation, periodically selling overweight assets and buying underweight ones. Wealth-management studies already envision using smart contracts to "automate portfolio rebalancing" when allocations drift beyond thresholds. GADE's system would coordinate these strategies: e.g. grid-trading or pairs-trading bots could be coded to run 24/7, adaptively setting buy/sell grids based on market volatility. In effect, the platform's AI-assistant continuously executes complex multi-asset strategies (dollar-cost averaging, momentum trading, hedging) that would be impractical to do manually. The net result is "sophisticated portfolio management" automated on-chain, removing emotional biases and enforcing disciplined risk control.

All of these features are exposed to ordinary users through **open APIs and intuitive dashboards**. Developers can tap GADE's REST and WebSocket APIs (similar to how current platforms do) to build client interfaces or trading bots. For example, today's developer-first brokers (like Alpaca) advertise easy-to-use REST APIs that let "developers and businesses trade algorithms, build apps and embed investing into their services". GADE would offer the same: a documented open API so any user, bot, or third-party app can programmatically access account data, market feeds, and order entry. In addition, GADE's web and mobile GUI would surface real-time analytics—charts, risk dashboards, trade simulation tools—so retail traders get "institutional-grade" insights. Think of a user logging in and instantly seeing cross-asset heat maps, factor analyses, and live AI signals alongside order forms. In this way GADE empowers everyday traders with the kind of automated tools and data that were once exclusive to quant funds.

11.8 Quantum-Secure Continuous Audit Infrastructure

A next component is an on-chain auditing system that assumes the coming quantum era and leverages AI monitoring. All GADE smart contracts and ledgers would use **post-quantum cryptography**. For instance, digital signatures on transactions and data could use lattice-based algorithms like CRYSTALS-Dilithium (known as ML-DSA) which NIST has standardized for post-quantum security. In practice, this means GADE's keys and signatures remain secure even against future quantum computers. All data at rest (historical ledgers, state roots) would be hashed/signed with similarly quantum-resistant schemes, effectively "hardening" the entire chain against any cryptographic break.

On top of this, GADE incorporates **continuous anomaly detection** powered by AI. Every transaction and financial event on the blockchain (plus off-chain ERP feeds, exchange records) is fed into a real-time risk engine. Machine learning models continuously scan for unusual patterns—fraudulent transfers, revenue spikes, or compliance exceptions. Research and practice in audit show AI can flag "anomalies, identify patterns, and assess risk more effectively" than manual methods. For example, if some ledger shows a sudden large sale to a flagged address, the system would alert auditors instantly. By contrast with traditional periodic audits, GADE's system means issues are caught in "near real-time" as soon as the data appears on-chain. In effect the platform provides a **continuous audit**: ML classifiers and neural nets monitor the entire transaction stream 24/7, automatically raising exceptions or even triggering lock-downs if something looks off.

Smart contracts also automate reconciliation. Audit rules (e.g. matching purchases to payments) can be coded into on-chain routines. For instance, a contract could require that total crypto withdrawals equal total cash deposits for a given day; if they don't, it flags a discrepancy immediately. Auditors no longer need to sample; as Deloitte observes, with blockchain "auditors can easily verify transactions on publicly available ledgers… [and] test the whole population of transactions…drastically improv[ing] the level of assurance". GADE extends this: every financial assertion (revenue, ownership, risk limits) is backed by on-chain proofs and automated checks. The platform can then **generate audit reports** that combine blockchain evidence with traditional GAAP/IFRS checks. For example, it might output a hybrid assurance report showing that assets are verified on-chain (PCAOB criterion: existence) while also noting any off-chain contracts needed for completeness. By design these reports would map to existing standards (IFRS, PCAOB, SOC, etc.), but they'd be filled with live data. In summary, GADE's audit layer enforces post-quantum security and uses AI to "provide continuous, real-time transaction auditing", making audits "faster, more comprehensive, and more predictive" than ever before.

11.9 Key Use Cases and Impact Metrics

GADE 2.0's combined features enable several transformative use cases:

- **Cross-asset Portfolio Trading and Rebalancing**: A trader can hold a portfolio of tokenized stocks, commodities, and crypto in one smart wallet. GADE's AI triggers will rebalance this portfolio in real time based on market moves and user-defined targets. For instance, if crypto rallies, the system automatically sells some crypto and buys back equity tokens to maintain target weights. This "live" multi-asset rebalancing unlocks strategies that today require juggling many platforms.
- **Retail-Trader Empowerment**: By democratizing the same tools used by Wall Street quants, GADE aims to reach **100 + million retail users** globally. Today exchanges like Coinbase and Binance each serve tens of millions of traders; GADE's vision is to combine crypto and traditional finance into one hub, offering institutional analytics (AI signals, portfolio stress tests) to retail users. In effect, over 130 million retail investors could gain free or low-cost access to quant analytics and automated strategies, shrinking the gap between retail and institutional capabilities.
- **Ultra-rapid Audits**: Thanks to on-chain settlement and continuous monitoring, full audits that once took months can be reduced to minutes. Instead of after-the-fact ledger reconciliations, auditors (or even regulators) can run on-demand queries of the GADE blockchain. For example, compliance officers could verify an exchange's entire transactions instantaneously, with AI explaining any anomalies. As Deloitte noted, blockchain enables "continuous online assessments" of all data, meaning the audit cycle time collapses from quarterly or annual sampling down to essentially real time.

Below is a summary of key performance indicators (KPIs) illustrating GADE 2.0's impact in quantitative terms:

KPI	Traditional benchmark	GADE 2.0 target
Settlement time	T + 2 trading days	Sub-second (instant)
Transaction fees	~ 0.2–0.5% per trade	~ 0.02–0.05% (≈90% reduction)
Liquidity/asset classes	Limited (crypto-only or stock-only)	Unified multi-asset pools (crypto, stocks, commodities, synthetics)
Audit coverage	~ 10–20% of transactions sampled	100% continuous (all transactions)

(continued)

(continued)

KPI	Traditional benchmark	GADE 2.0 target
Signal accuracy	~ 60–70% (manual/heuristic)	~ 90% (AI-ensemble predictions)
Audit cycle time	Weeks–months	Minutes (real-time audit)
Retail users served	Tens of millions per platform	> 130 million (combined reach)

Each of these reflects orders-of-magnitude improvements. For example, sub-second settlement and near-zero fees via stablecoin rails significantly reduce the typical settlement delays and costs of fiat markets. Automatic, AI-driven monitoring raises audit coverage from a sampling fraction to full coverage. And by offering open APIs and user-friendly dashboards, GADE can bring trading and analysis to a vast retail audience as easily as existing consumer Fintech apps, but with institutional-grade depth.

Looking ahead, GADE 2.0's innovations bridge the worlds of crypto and traditional finance. By tokenizing real-world assets, unifying cross-asset liquidity, and embedding AI and quantum-secure audit tools, it paves the way for a truly integrated financial markets ecosystem. In the final analysis, GADE illustrates how cutting-edge blockchain, AI, and cryptography can converge to create faster, cheaper, more transparent markets—a fitting lead—in to the book's concluding discussion of the future of finance.

11.10 Conclusion and Future Outlook

The vision of a unified, tokenized, AI-automated, and quantum-resilient financial ecosystem is now within reach. By weaving together innovations like GADE 2.0, the AI-driven hub, and a continuous audit layer, we imagine a single coherent infrastructure where every asset, service, and compliance check flows seamlessly. Physical and digital assets become native tokens, enabling instant, borderless value transfer. Intelligence is embedded at every layer: automated market-making, dynamic risk assessment, and real-time compliance adapt on the fly. Quantum-resistant cryptography underpins every transaction, ensuring that this future is secure even against tomorrow's computing breakthroughs. In this ecosystem, finance is not fragmented silos but a harmonious whole—smarter, faster, and fundamentally more trustworthy.

11.11 2030 and Beyond: The New Financial Norms

Looking ahead to 2030 and beyond, the financial landscape will feel transformed. What once were cutting-edge ideas have become everyday realities:

- **Autonomous Markets**: Trading, settlement, and regulatory compliance are largely automated. Smart contracts negotiate prices and execute transactions instantly, without manual intervention. By 2030, "trade-and-forget" processes are standard; asset exchanges happen in milliseconds with built-in, AI-monitored compliance checks at every step.
- **AI-Driven Asset Management**: Advanced AI advisors and portfolio managers are widely accessible to institutions and retail investors alike. Personalized, algorithmic asset strategies powered by GADE 2.0's intelligence adapt in real time to market shifts, far outperforming static portfolios. Wealth management becomes as easy as speaking to a personal digital advisor, democratizing sophisticated strategies once reserved for the elite.
- **Quantum-Ready Security**: Post-quantum cryptography is baked into all financial infrastructure. From quantum-resistant ledgers to biometric authentication secured by lattice-based encryption, every layer anticipates future threats. Institutions actively rotate to quantum-safe algorithms, and legacy financial records are rekeyed to withstand quantum attacks. In this era, trust in the system grows because the weakest link—security—has been future-proofed.

As these norms solidify, the line between traditional finance and crypto-finance blurs completely. Central banks, banks, Fintechs, and DeFi platforms all operate on shared rails. The result is a robust, global financial fabric: fast, borderless transactions; ubiquitous personalized finance; and ironclad trust rooted in cryptography.

11.12 Emerging Frontiers in Finance

Beyond these foundations, entirely new frontiers emerge to expand possibilities:

- **Composable DeFi and Modular Primitives**: Finance is built like Lego. Open building blocks (lending pools, AMMs, insurance primitives, Oracle networks, and more) interlock seamlessly across platforms. Developers assemble bespoke financial products by combining standardized modules, fuelling rapid innovation. Need a new yield strategy? Simply plug together the right lending, staking, and insurance bricks. These modular primitives make the ecosystem infinitely extensible and user-driven.
- **Decentralized Identity and Self-sovereign KYC**: Individuals own and control their financial identities. Self-sovereign ID systems allow users to prove credentials (credit history, accreditation, compliance status) without centralized intermediaries. On-boarding to any financial service is as simple as unlocking a digital wallet—no repetitive paperwork or privacy compromises. Regulators and banks recognize universal identity standards on-chain, making KYC interoperable and transparent. Privacy-enhancing technologies ensure data is shared only with consent, securing user autonomy.

- **Cross-chain Interoperability as the Norm**: Multiple blockchains and ledgers coexist and interconnect fluidly. Assets and data flow freely between networks via standardized bridges and protocols. No more siloed ecosystems—whether it's CBDCs, bitcoin, Ethereum-based assets, or private networks, funds move seamlessly and securely across chains. Global finance leverages this interconnected mesh: a payment on one chain automatically triggers settlements or compliance checks on another. This true interoperability fuels a global market where liquidity and opportunities are boundless.

Each of these frontiers represents a paradigm shift. Together, they create an open, programmable financial universe where anyone—from a solo entrepreneur to a multinational—can participate in any service, any geography, in real time.

11.13 Policy and Industry Shifts for a New Era

Realizing this future also requires deliberate changes in policy and industry practices:

- **Collaborative Regulatory Sandboxes**: Governments, industry, and academia set up global innovation hubs. In these sandboxes, regulators experiment alongside start-ups with new financial products under real-world conditions. This collaborative approach accelerates safe innovation—testing everything from tokenized securities to AI-driven compliance tools—while regulators learn and adapt in lockstep. By breaking down silos between policy and practice, sandboxes nurture trust and ensure regulation follows, not lags, innovation.
- **Standardization of Token Formats and Audit Trails**: Industry consortia and international bodies define common token standards and audit protocols. Whether representing equities, bonds, or real-world assets, tokens share a universal format. This standardization simplifies cross-market trading and clearing. In parallel, audit trails become automatic and encrypted: every transaction attaches tamper-proof metadata for identity, provenance, and compliance. Auditors and regulators can access real-time ledgers that speak the same language worldwide.
- **Open-Data and AI Model Sharing Frameworks**: Data and intelligence become public goods. Consortia release open financial datasets and AI models for analytics, risk scoring, and fraud detection. Start-ups and smaller institutions tap into these shared resources to level the playing field with industry giants. At the same time, open frameworks ensure model transparency and fairness. This democratization of data and AI tools accelerates innovation and drives performance across the board.
- **Encouraging Post-quantum Readiness**: Policymakers incentivize quantum-safe transitions. Global standards bodies mandate quantum-resistant encryption for critical financial infrastructure by a set deadline. Industry initiatives help smaller players migrate legacy systems to new algorithms. Governments coordinate cross-border risk assessments for quantum threats. This proactive stance means the

entire financial ecosystem stays ahead of quantum computing advances, rather than scrambling reactively.

These shifts create an ecosystem where innovation is responsibly encouraged and widely shared. Regulators become partners in progress, communities co-develop standards, and public institutions fund open platforms. The result is a level playing field where both incumbents and newcomers thrive under clear, future-proofed rules.

11.14 Final Thoughts

The journey to a fully unified, AI-empowered, and quantum-secure financial system is ambitious, but within our grasp. By 2030 and beyond, the fusion of blockchain tokenization, advanced AI, and robust cryptography will revolutionize how value flows around the world. In this future, markets operate 24/7 with zero friction, financial services adapt instantly to our needs, and every transaction carries built-in trust and transparency.

None of this is science fiction—it is a roadmap drawn by today's pioneers. As we turn this vision into reality, collective effort will be key. Financial institutions, technologists, policymakers, and citizens must all play their part. With collaborative governance, shared innovation, and a steadfast commitment to security and openness, the coming decade will usher in an era of unparalleled prosperity and inclusion. The infrastructure we have designed is not just about cutting-edge technology; it is about reshaping finance to serve humanity better. By embracing this future together, we can ensure that by 2030 and beyond, global finance truly serves everyone, everywhere, as a unified, intelligent, and secure system.

References

1. https://www.weforum.org/publications/global-retail-investor-outlook-2025/?utm_
2. https://www.frontiersin.org/journals/computer-science/articles
3. https://dblp.org/rec/journals/corr/abs-2503-21422.html?utm_
4. https://www.sidley.com/en/insights/newsupdates/2024/12/artificial-intelligence-in-financial-markets-systemic-risk-and-market-abuse-concerns?utm_
5. https://dblp.org/rec/journals/corr/abs-2503-21422.html?utm_
6. https://www.sidley.com/en/insights/newsupdates/2024/12/artificial-intelligence-in-financial-markets-systemic-risk-and-market-abuse-concerns?utm_
7. https://web3unplugged.io/wp-content/uploads/2024/10/From-ripples-to-waves-The-transformational-power-of-tokenizing-assets.pdf?utm_
8. https://finukr.com.ua/index.php/journal/article/view/82?utm_
9. https://ijrpr.com/uploads/V6ISSUE5/IJRPR45370.pdf?utm_
10. https://www.deloitte.com/mt/en/services/audit-assurance/perspectives/mt-blockchain-a-game-changer-for-audit.htmlv
11. https://www.lseg.com/content/dam/data-analytics/en_us/documents/charts/lseg-size-of-global-market-2024-in-charts.pdf

12. https://www.bain.com/insights/ai-in-financial-services-survey-shows-productivity-gains-across-the-board/
13. https://www.bis.org/cpmi/publ/swift_gpi.htm
14. https://www.pwc.com/gx/en/services/audit-assurance/internal-audit/global-internal-audit-study.htm
15. https://www.mckinsey.com/industries/financial-services/our-insights/global-payments-in-2024-simpler-interfaces-complex-reality
16. https://www.bis.org/cpmi/publ/swift_gpi.htm#:~:text=%2A%20The%20speed%20of%20cross,to%20occur%20in%20low%20and
17. https://www.mckinsey.com/industries/financial-services/our-insights/global-payments-in-2024-simpler-interfaces-complex-reality?utm_
18. https://www.bobsguide.com/global-payments-2024-the-hidden-complexity-of-simpler-payments/?utm_
19. https://www.paynet.pro/post/how-to-make-the-most-of-the-next-1-trillion-growth-in-global-payments?utm_
20. https://www.paymentscardsandmobile.com/global-payments-2024-simplicity-but-complexity-beneath/?utm

Index

P. M. Jeyanthi (ed.), *Distributed Ledger Technology in Communication*, Signals and Communication Technology, https://doi.org/10.1007/978-3-032-07682-3